Sustainable Textiles: Production, Processing, Manufacturing & Chemistry

Series Editor

Subramanian Senthilkannan Muthu, Chief Sustainability Officer
Green Story Inc.
Tung Chung, Hong Kong

This series aims to address all issues related to sustainability through the lifecycles of textiles from manufacturing to consumer behavior through sustainable disposal. Potential topics include but are not limited to: Environmental Footprints of Textile manufacturing; Environmental Life Cycle Assessment of Textile production; Environmental impact models of Textiles and Clothing Supply Chain; Clothing Supply Chain Sustainability; Carbon, energy and water footprints of textile products and in the clothing manufacturing chain; Functional life and reusability of textile products; Biodegradable textile products and the assessment of biodegradability; Waste management in textile industry; Pollution abatement in textile sector; Recycled textile materials and the evaluation of recycling; Consumer behavior in Sustainable Textiles; Eco-design in Clothing & Apparels; Sustainable polymers & fibers in Textiles; Sustainable waste water treatments in Textile manufacturing; Sustainable Textile Chemicals in Textile manufacturing. Innovative fibres, processes, methods and technologies for Sustainable textiles; Development of sustainable, eco-friendly textile products and processes; Environmental standards for textile industry; Modelling of environmental impacts of textile products; Green Chemistry, clean technology and their applications to textiles and clothing sector; Eco-production of Apparels, Energy and Water Efficient textiles. Sustainable Smart textiles & polymers, Sustainable Nano fibers and Textiles; Sustainable Innovations in Textile Chemistry & Manufacturing; Circular Economy, Advances in Sustainable Textiles Manufacturing; Sustainable Luxury & Craftsmanship; Zero Waste Textiles.

Sadhna • Rajesh Kumar • E Mokanaasri
Editors

Crafting Sustainability in Luxury Textiles for a Zero-Waste Future

Springer

Editors
Sadhna
Department of Fashion Design
Woxsen University
Hyderabad, Telangana, India

Rajesh Kumar
Department of Fashion Design
Woxsen University
Hyderabad, Telangana, India

E Mokanaasri
Department of Fashion Design
Woxsen University
Hyderabad, Telangana, India

ISSN 2662-7108 ISSN 2662-7116 (electronic)
Sustainable Textiles: Production, Processing, Manufacturing & Chemistry

ISBN 978-3-031-86772-9 ISBN 978-3-031-86773-6 (eBook)
https://doi.org/10.1007/978-3-031-86773-6

This Springer imprint is published by the registered company Springer Nature Switzerland AG
The registered company address is: Gewerbestrasse 11, 6330 Cham, Switzerland

Preface

The luxury textile industry is essential in a world increasingly defined by environmental challenges and the urgent need for sustainable practices. Historically associated with opulence, exclusivity, and artisanal craftsmanship, luxury textiles have often been regarded as a symbol of timeless beauty and unparalleled quality. However, as the global call for sustainability intensifies, the industry must reconcile its heritage of luxury with the imperative of environmental responsibility. *Crafting Sustainability in Luxury Textiles for a Zero-Waste Future* is a timely and essential contribution to this transformative dialogue.

This edited volume explores the intersection of luxury and sustainability, offering a nuanced perspective on how the two can coexist harmoniously in the textile sector. By drawing on a rich diversity of voices, including academics, designers, artisans, and industry leaders, the book comprehensively examines the challenges and opportunities that define this crucial shift. It illuminates how sustainability can be integrated into every aspect of luxury textile production—from raw material sourcing and design innovation to manufacturing processes and consumer engagement—without compromising the essence of luxury.

One of this book's key strengths is its focus on zero waste. By addressing the wasteful practices often associated with the textile industry, the contributors underscore the importance of resource efficiency, circular design principles, and waste minimization strategies. These principles are critical to achieving sustainability and aligning seamlessly with the ethos of luxury, which values precision, craftsmanship, and thoughtful use of resources.

Equally significant is the book's emphasis on craftsmanship and cultural heritage. Luxury textiles have long been a repository of traditional knowledge, artisanal skill, and cultural narratives. As the industry embraces sustainability, it has a unique opportunity to celebrate and preserve these traditions while fostering innovation. The contributors highlight inspiring examples of how artisans and designers are reimagining luxury through sustainable practices, creating textiles that are both environmentally responsible and deeply rooted in cultural authenticity.

This volume also addresses the broader implications of sustainability in the luxury textile sector, including its economic, social, and ethical dimensions. The chapters delve into topics such as fair labor practices, community empowerment, and role of consumer awareness in driving sustainable choices. By adopting a holistic approach, the book advances our understanding of sustainable luxury textiles and provides actionable insights for stakeholders across the value chain.

As we navigate an era of unprecedented change, *Crafting Sustainability in Luxury Textiles for a Zero-Waste Future* serves as both a reflection of current progress and a road map for future innovation. It challenges us to envision a textile industry where luxury is defined not by excess but by its ability to inspire, endure, and contribute positively to the planet. This book is a call to action for all those invested in the future of textiles, urging us to embrace sustainability as a core principle rather than a mere trend.

This volume will resonate with a broad audience, including researchers, practitioners, educators, and consumers. Its rich blend of academic rigor and practical insights makes it an invaluable resource for anyone seeking to understand and contribute to the evolution of luxury textiles in the context of sustainability. I commend the editors and contributors for their visionary work in creating a book that is both intellectually stimulating and profoundly relevant to the challenges of our time.

Hyderabad, Telangana, India Sadhna

Contents

Acknowledgments

It is with deep gratitude and appreciation that I acknowledge the many individuals and organizations who have contributed to the development of *Crafting Sustainability in Luxury Textiles for a Zero-Waste Future*. This book is the product of collective effort, intellectual collaboration, and a shared passion for sustainability and innovation in the textile industry.

First and foremost, I extend my heartfelt thanks to the contributors whose scholarly expertise, insights, and dedication have been central to the creation of this volume. Your chapters have provided a wealth of knowledge and have enriched the discourse on sustainability, luxury, and textile innovation. This book would not have been possible without your valuable contributions.

I am profoundly grateful to the editorial team, Rajesh Kumar and E Mokanaasri, for their tireless efforts, guidance, and unwavering support throughout this journey.

A special note of thanks goes to the reviewers who provided thoughtful critiques and valuable suggestions during the development process. Your input has significantly enhanced the academic rigor and depth of this volume.

I would like to acknowledge the Springer publishing team for their expertise and professionalism in bringing this project to fruition. Your support, attention to detail, and commitment to quality have been invaluable.

I sincerely appreciate the artisans, designers, and industry professionals whose work and dedication continue to inspire the themes explored in this book. Your commitment to integrating sustainability into luxury textiles is a testament to the design's transformative potential.

Lastly, I wish to thank the readers of this book, who share a passion for sustainability and the future of luxury textiles. Your engagement with this work contributes to the growing direction toward a more ethical and environmentally responsible industry.

Crafting Sustainability in Luxury Textiles for a Zero-Waste Future is dedicated to all those who envision a world where luxury and sustainability coexist harmoniously. It is my hope that this book serves as a motivation for innovation, dialogue, and positive change in the textile industry and beyond.

Sadhna

Introduction: Evolution of Sustainable Textiles

Sadhna, E Mokanaasri, Rajesh Kumar, and S. Greeshma

1 Introduction

Textiles have come a long way from being a simple by-product of the historic local cottage industry to the miserable fast fashion system that exists worldwide today. This evolution has not been without its price: the exploitation of natural resources, the quality of the environment such as polluted water, depleted resources, and increased waste production to the highest level of textile production. Textiles in the past were produced from natural fibers and natural dyes that are reckoned to have a less impact on the environment than the current methods. However, Fig. 1 shows that the industrial revolution into mechanization revolutionized the ability to produce rapidly, but at the same time, it raised many widespread environmental issues such as amplified water pollution and consumption of resources [1].

Paying attention to the environmental issues linked to conventional processes of textile production, the number of sustainable initiatives has recently increased in the industry. The participation of ecological fibers like organic cotton, hemp fiber, and recycled fiber also disseminates the impacts of conventional fiber. These sustainable substitutes are meant to reduce the negative impact on the environment as much as to satisfy the customer's needs for quality and longevity [2]. Moreover, increased use of the label and certification about the ethical and ecological images of textile products has encouraged consumers to demand more sustainable textiles [3]. However, while there has been progress in the integration of sustainability into the textile industry development, several challenges are still realized in the large-scale implementation of sustainability. The fast fashion model is unsustainable with very poor environmental management; every time, there has been extensive consumption

Sadhna · E. Mokanaasri (✉) · R. Kumar · S. Greeshma
Department of Fashion Design, School of Arts and Design, Woxsen University, Hyderabad, Telangana, India

Sadhna et al. (eds.), *Crafting Sustainability in Luxury Textiles for a Zero-Waste Future*, Sustainable Textiles: Production, Processing, Manufacturing & Chemistry, https://doi.org/10.1007/978-3-031-86773-6_1

Fig. 1 Evolution of luxury textiles

and discard of textile waste, and it is estimated that 85% of clothes that are unwanted go to the dumps [4].

Further, since sustainable materials cost more and recycling requires additional expenses, not every company is in a position to incorporate environmental concerns in their manufacturing processes [5]. Today, consumers show a tendency to research the origin of the clothes they buy and make a purchase only if the materials and manufacturing process meet their expectations of sustainability, so luxury brands have been coming up with modern unique solutions. That being said, luxury and sustainability are not entirely symbiotic, or without their risks for the textile industry as it strives for profit and planet preservation [6].

2 Historical Context of Sustainable Textiles

The history of sustainable textiles began with the ancient ways when cotton, wool, silk, and flax were all used, and these fibers were produced on-site or locally and then processed without much impact on the environment. Methods for dyeing and weaving are resource-friendly, developed under the principles of scarcity instead of plenty. Mass production during the Industrial Revolution gave way to synthetic fibers and the destruction of the environment. In response, this modern sustainability movement aims to revive ecofriendly practices while incorporating technological advancements to reduce the ecological footprint of textile production.

2.1 The Origins of Textile Production

The evolution of the textile industry can convincingly be traced back to the earliest civilization starting with the Mesopotamian, Egyptian, and Chinese civilizations. Textiles were made from natural fibers such as cotton, wool, and silk, and created using craft techniques, and often the spinning and weaving were done by hand. These practices were normally bench-scale, and thus, their impacts on the environment were relatively small compared to current commercial methods.

Conventional practices in the production of textiles have been established to preserve the environment compared to current practices. Handmade products were obtained from natural materials, and local resources and used natural dyes that had less effects on the environment. Even these practices were further associated with some problems. It was determined that the emission of pesticides used in cotton cultivation and water consumption in the dying of fabrics were the factors that led to the pollution of the local environment and depletion of resources. For example, traditional methods of growing cotton require approximately 6% of world pesticide use, including 16% of worldwide insecticides, while harming the immediate environment [1, 7].

The Industrial Revolution therefore literally brought a revolution to the production of textiles and mechanization, which in turn brought the mass production capacity that in turn set very high production rates but at the same time had very negative effects on the environment. This led to massive pollution of water—dyeing and finishing were estimated to be responsible for 20% of the world's clean water pollution. Generally, the fashion industry is one of the heaviest consumers of freshwater in the world and contributes to about 10% of industrial water consumption. It takes approximately 2700 L of water to make a single cotton t-shirt and a personal water consumption of one person for 2.5 years of drinking water. The effects of this high water consumption include urgent hitches in the local water supply, especially where water is already a scarce resource [8].

Mirror societal changes smaller alterations as seen before might occur in the future textiles are evidence of societal changes. Starting from the hand-woven fabrics to the modern-day fast fashion model with cycles of fast production and fast consumption—each has its own set of environmental implications.

- **Ancient methods:** Traditional fabrics were environmentally friendly to come with fewer leftovers from locally obtained raw materials.
- **Industrial Revolution:** Complications profited on polluting the environment; this is evident whereby mechanization led to increased pollution, both air and water through chemical dyes and large-scale production.
- **Fast fashion era:** The last decades of the twentieth century became a time of a great degree of clothing consumption encouraged by fast fashion brands that implied intensive use of cheap resources. This has led to high waste chucks; for example, less than 1% of used clothing are reused to make new clothes [9].

2.2 The Rise of Sustainable Practices

Sustainable fiber and material development has, therefore, been spurred on by the rising concern of the environment within textile production. While the harms inflicted by conventional textiles on the environment were gaining increased attention, the sector has started to seek options that have the least negative effects on the environment. Sustainable fibers can be categorized into three main types:

- **Natural fibers:** These are the organic cotton, hemp, linen, and bamboo products that are grown using naturally derived pesticides or fertilizers. Organic cotton, for example, is grown in ways that preserve the health of the soil as well as use less water compared to normal growing conditions for most cotton crops [2].
- **Recycled fibers:** Fabrics like recycled polyester (rPET) have become more common because they use plastic scraps, thus freeing the industry from using fresh petroleum sources. Recycled fibers involve less energy than virgin fibers, and the texts also indicate that recycled fibers have less carbon footprint.
- **Innovative synthetic fibers:** Lyocell (Tencel) and polylactic acid (PLA) are relatively recent materials that have been developed from renewable sources with the intention that they will break down or decompose. These fibers are designed to ensure productivity while at the same time less harm to the environment [3].

Figure 2 shows a systematic flow of the textile and garment manufacturing process, from the procurement of raw fibers such as cotton, wool, silk, or synthetic materials. This material is further collected, processed, and wound into threads, then

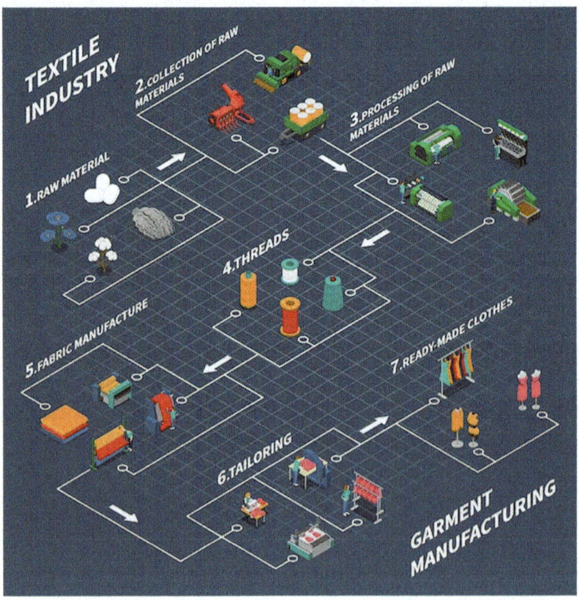

Fig. 2 Pathway from the textile industry to garment manufacturing [10]

woven or knitted to create fabrics. The fabrics are then subjected to tailoring, cutting, sewn, and joining together to form garments. Lastly, the ready-made clothes are manufactured and prepared for dispersal. It is all integrated back to the beginning stages.

The evolution of sustainable textiles has been marked by several key milestones since the 1960s:

- **1960s–1970s: Early environmental awareness:** Pressure groups for the environment emerged early in the 1960s with the advent of books such as Rachel Carson's "Silent Spring" on pollution. During this period, people started to talk about sustainability in different businesses such as textiles [11].
- **1980s: Eco-friendly material adoption:** The use of organic cotton in the manufacturing process began in the 1980/1990s. Industry leaders started to look for more natural fibers that are produced with less or no use of toxic chemicals in the production process [2].
- **1990s: Rise of eco-labels:** In the 1990s, there were developments in the so-called money-value-eco-labels like the Global Organic Textile Standard (GOTS) or the OEKO-TEX certification where the consumers could reassure themselves concerning the environmental and social conditions of the production of the cloth [3].
- **2000s: Mainstream adoption of sustainability:** The glamor for green products was felt at the close of the twentieth century and the beginning of the new millennium. Substantial food manufacturers started incorporating recyclable materials into those supply chains, along with extensively reporting on their product sourcing. Another aspect related to the fast popularity and production of clothing was criticized for the harm that it causes the environment, which means that people need to consume in a better way [4].
- **2010s: Innovations in sustainable textiles:** This decade marked improvements in textile technologies such as the production of biodegradable synthetics and dyeing through waterless systems. Consumers were witnessing the emergence of new types of materials invested in brands' Research and Development Department to improve the image or reduce the company's impact on the environment.
- **2020s: Circular economy focus:** At the present time, circular economy schemes are most actively discussed within the textile industry. There are also many rental services, secondhand websites and apps that either rent or resell clothes and apparel, and recycling outlets that try to reduce the amount of clothing purchased and thrown away. It also opens up more opportunities for consumers to receive and act upon information about sustainability.

2.3 *Sustainability Certifications and Standards in Textile Production*

In the last few decades, many certifications and standards within the textile industry have proliferated, and more attention has been paid to sustainability. Standards including the Global Organic Textile Standard (GOTS) and Fair-Trade certifications have helped act as standards for sustainability in textile production, giving consumers realistic information regarding the sustainability of textile products.

2.3.1 Global Organic Textile Standard (GOTS)

At present, the chief processing standard for textiles produced from organic materials is the Global Organic Textile Standard (GOTS). A global standard for organic textile production, GOTS was developed by a group of international organizations in organic agriculture to set ecological criteria and social responsibility in the global chain of affairs. This certification confirms that production materials starting from, harvesting of raw materials right from the manufacturing processes, labeling, and distribution of products conform to certain approved standards [12, 13].

- **Organic fiber content:** The requirements for GOTS certification state that a textile product must be made out of at least 70% organic fibers. For foods, beverages, and other products that have been marked as "organic," this provision rises to at least 95% organic materials.
- **Ecological criteria:** GOTS in its current version requires that all the chemicals used in processing must be environmentally and toxicologically certified. This entails banning such compounds and setting performance standards for the treatment of wastewater in the processing plants.
- **Social criteria:** GOTS also contains social responsibility standards that are related to those of ILO. These criteria include the provision of decent wages to workers, physical protection to people who work, a ban on the use of children in production, and the ability of workers to self-organize [14].
- **Third-party certification:** GOTS, therefore, entails third-party certification augmented by strict auditing by an authorized certification organization, hence offering closed admissions of the entire supply chain [15].

3 Current Challenges in the Textile Industry

Challenges in the textile industry include environmental issues resulting from high water consumption, pollution due to dye deposition, and textile waste. Many areas continue to exploit labor and have dangerous working conditions. Furthermore, the demand for fast fashion creates pressure on supply chains and increases overproduction, leading to unsustainable business models as shown in Fig. 3. The biggest challenge that the industry has to face is balancing profitability with sustainability and ethical considerations.

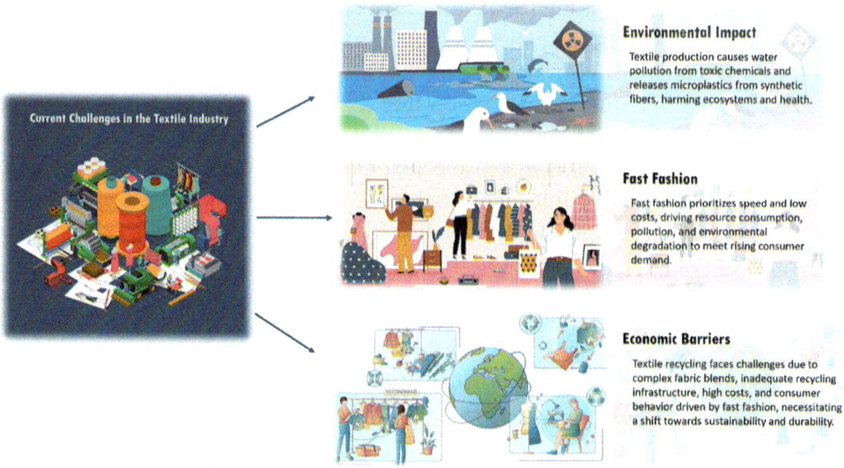

Fig. 3 Current challenges in the textile industry [16–18]

3.1 Environmental Impact

This details the textile industry's contributions to pollution, including water usage, chemical discharge, and waste management issues, and highlights the scale of these challenges on a global level.

- **Chemical discharge:** Two essential stages in the production of textiles, which are dyeing and finishing, are significant producers of water pollution. Others estimate that the textile production sector is responsible for approximately 20% of industrial water pollution worldwide, caused by the dumping of chemical substances into water. This chapter finds that most factories across the developing countries do not have adequate water treatment systems; hence, most of the released wastewater goes directly to the rivers and oceans, polluting water and posing threats to water users.
- **Microplastic pollution:** Polyester as a form of synthetic fiber has presented distinct impacts on the environment that comprise microplastics type. Every time synthetic apparel is washed, the microfibers are shed—very small and capable of entering the marine systems. Each laundry wash releases an estimated 700,000 microplastic fibers to the ocean, which is currently struggling to contain and manage plastic waste. These microplastics not only affect marine wildlife but also become incorporated into the food chain to present threats to people's health [4].
- **Overconsumption and disposal:** Fast fashion predisposes consumers to waste by providing them with clothes that are made quickly at minimal expense to suit the latest fads. Therefore, consumers chuck out roughly 85% of textiles annually; the majority are buried in landfills or subjected to incineration. The tragedy is

that only about 50% of the used clothes are collected for reuse or recycling and as of 2007 only 1% of the collected clothes were recycled into new wear.

- **Waste management issues:** The fact that new trends come up frequently contributes to high production of textile waste. Each year, a European citizen sends approximately 11 kg of textile to the landfill, increasing the problem of waste. The problem posed by this increase is that most countries just do not have sufficient waste disposal facilities that can handle this stream, and this leads to environmental pollution, and the health of communities living near these dumpsites [16].

3.2 Fast Fashion Has Fundamentally Altered Consumer Behavior and Industry Practices

- **Production cycles:** The current industry model is fast fashion, which entails faster cycles, higher turnover rates, and little regard for quality and sustainability. It is evident that Zara or H&M or any other fast fashion clothing brand is incapable of producing mere weeks' worth of clothes. This not only leads to excessive utilization of resources and the generation of greater pollution at each step from sourcing of raw materials to the production and delivery of the final product.
- **Market demand:** The greater pressure consumers put on clothing brands for affordable and fashionable products, the greater pressure that clothing brands place on manufacturers to use cheap methods that have enormous negative impacts on the environment and social justice. Such a cycle institutionalized the process of externalization of ecological consequences, thus worsening the environmental situation globally, although maintaining the lowest possible prices for consumers [17].

3.3 Economic Barriers

Economic viability: Recycling textiles remains economically sustainable, which is one of the biggest factors affecting the process. We know that recycling processes can sometimes be expensive to undertake because of the technologies and human resources required. For instance, in mechanical recycling, the fibers are made weaker to have low quality and cannot be used again. Furthermore, the recycling cost of textile material is even higher than that of virgin material means that the garment manufacturers do not have adequate incentives to recycle [18].

Consumer behavior: Fast fashion contributes to the formation of the specified unsustainable pattern of consumption because the process is driven by discarding clothes in large numbers frequently. People wear garments for only a few uses before discarding them, creating large volumes of textiles that far outstrip the presently existing recycling capacity. Consumer behavioral shift toward perceiving durability and sustainability as valuable is crucial for the promotion of a circular economy [19].

4 Intersection of Luxury and Sustainability

Fashion giants such as Gucci and Prada have begun acting ethically when sourcing fabrics; using organic and recycled fibers. They value transparency and other brands such as Kering apply transparency in the supply chain. Eco-friendly and sustainable processes cut wastage; simple and elegant designs, focus on quality, luxury reducing the environmental footprint, and reusing materials, like Valentino's deadstock apparel.

(a) **Luxury brands are advancing sustainability:** Cultures of purpose: Gucci aims for its net-zero emission and material renewal while Prada's "Re-Nylon" is the use of regenerative nylon instead of virgin ones, Stella McCartney eliminates the use of animal products, Chanel announces its "Mission 1.5" that wants to reduce its carbon emission by 50% by the year 2030 while exploring the use of regenerative agriculture and supply chain transparency [6]. Consumers wish more companies to be clear about their supply chain and production; consumers are choosing quality, long-lasting products over mass-produced cheap products. They are committed to well-made and long-lasting and are in line with sustainable luxury. Many consumers are also willing to pay a premium for brands that are more conscious of their ethical and sustainability responsibilities, thereby offering poor brands an excuse to overcharge for their products [20].

(b) **Challenges for luxury brands:** Regulations vary across the globe; thus, it is difficult to manage a sustainable supply chain. Sustainable supply chain management proved to be difficult due to different regulations across the world. Consumers expect products of high quality and sustainably sourced materials; however, new materials may be hard to sustain. They noted that the cost of going green reduces profits because the related expenses are normally costly. As more players have entered into the sustainable luxury product sector, luxury brands need to set themselves apart while focusing on sustainability [21].

5 Pioneering a Zero-Waste Future in Textile Crafting

Zero-waste design in the textile industry works toward preventing wastage during a product's life cycle through proper use of material, unique and innovative cutting techniques, and designing a product for disassembly. This approach fosters a circular economy because it optimizes the use of resources due to reuse, recycling, and recovery practices that minimize the output of wastes into environments [22].

5.1 *Examples of Brands Implementing Circular Fashion Models*

Eileen Fisher's Renew Program retracing and upcycling: It prevents wastage through opportunities for multiple cycles of use of materials. Zero Waste Daniel produces fashionable apparel from preindustrial waste, meaning fabrics from the New York City garment industry waste products. Hellen Van Rees designs her wearables without waste and uses offcuts of yarns and such methods as knotting, fusing, and knitting [23].

5.2 *Exploration of New Sustainable Materials (e.g., Mycelium Leather, Recycled Fabrics)*

Mycelium leather, derived from the root structure of mushrooms, is emerging as a sustainable alternative to animal leather, with companies like Bolt Threads creating biodegradable materials that mimic leather's properties. Additionally, recycled fabrics, such as polyester made from post-consumer plastic bottles, are gaining traction in the textile industry, with brands like Patagonia and Adidas utilizing recycled materials to lessen the demand for virgin resources [24].

5.3 *Pioneering a Zero-Waste Future*

- **Case studies of zero-waste initiatives:** Zero Waste Daniel is a fashion project that was created by Daniel Silverstein, garments made from pre-consumer waste, collected locally, from the NYC garment industry, with no-waste cutting techniques. The main collection has created spectacular collections, motivating environmentally friendly people through converting cuttings into chic dresses. It is a big problem to provide sufficient waste material supply for production needs meeting the qualitative, aesthetic, and functional expectations [25].
- **Collaborations and partnerships:** It is a global platform that unites brands, manufacturers, NGOs, and universities to advance sustainable fashion practices. For instance, the "Good Fashion Fund" has already been presented to startups developing circularity. "Make Fashion Circular" is the CR model by the Ellen MacArthur Foundation to spread the circular economy by teaming up with leading fashion players. Recycling, reuse, and sustainable design in the textile sector receive a boost from programs like the extended producer responsibility (EPR) schemes promoted by the EU government [26, 27].

6 Conclusion

Several changes have occurred in the past few decades toward attaining sustainable textiles. Historically, the fashion industry was associated with simple, through-the-line production and consumption that caused a significant amount of pollutive impacts on the environment and socially unjust effects. But rising consciousness has led to a shift toward sustainability fostered by new materials, design techniques, and customers' preferences of today's sustainable textiles including the use of environmentally friendly fibers, minimum waste design, and the circular fashion system. To improve sustainable standards of the fashion industry, collaborations between brands, governments, and NGOs have become important driving forces and alliances such as The Fashion Pact. Consequently, it is the preponderance of new concepts and initiatives related to transparency, selectivity in sourcing materials, and proper consumption among fashion producers and customers.

Luxury and sustainability is an undiscovered concept that, when considered, can help completely rethink luxury and turn it into craftsmanship, nonecological impact, and responsible sourcing. The sustainability revolution has started already with the most luxurious brands putting into practice more sustainable procedures and strategies that other brands should follow. With transparency, sustainable investment, and active consumer dialogue, luxury asserts its influence to take sustainability throughout the textile supply chain of the fashion industry.

In light of these transitions, the sustainability of the textile industry should become a fundamental goal for all related parties. Manufacturers must be more responsible, designers have to think about the cradle-to-cradle approach, and consumers should only buy from environmentally conscious companies. With research as a key driving force and through a dedication to green innovation, the textile industry has an opportunity to demonstrate how luxury and sustainability can go hand in hand.

References

1. Roy Choudhury, A. K. (2014). Environmental impacts of the textile industry and its assessment through life cycle assessment. In S. S. Muthu (Ed.), *Roadmap to sustainable textiles and clothing* (pp. 1–39). Springer. https://doi.org/10.1007/978-981-287-110-7_1
2. Ali, M. A., & Sarwar, M. I. (2010). *Sustainable and environmental friendly fibers in textile fashion: A study of organic cotton and bamboo fibers* (University of Borås; Master Thesis 14 September 2010, University of Borås). https://www.divaportal.org/smash/get/diva2:1312071/FULLTEXT01.pdf
3. Shah, D. (2023, May 15). *Sustainable fibres for a greener fashion industry*. Fibre2Fashion. https://www.fibre2fashion.com/industry-article/9636/sustainable-fibres-for-a-greener-fashion-industry
4. Šajn, N., & European Parliamentary Research Service. (2019). Environmental impact of the textile and clothing industry. In *European Parliamentary Research Service (EPRS Report PE 633.143)*. https://www.europarl.europa.eu/RegData/etudes/BRIE/2019/633143/EPRS_BRI(2019)633143_EN.pdf

5. Emmanuel, J. (2024, February 19). *Fashion's eco-challenge: Scaling the textile recycling process*. Woven Insights Blog. https://woveninsights.ai/site-blog/fashions-eco-challenge-scaling-the-textile-recycling-process/
6. Birch, K. (2024, February 19). Top 10: Sustainability strategies of luxury fashion brands. *Sustainability Magazine*. https://sustainabilitymag.com/sustainability/top-10-sustainability-strategies-of-luxury-fashion-brands
7. Prehu, I. (2024, April 2). 5 environmental impacts of textile industry, and 3 ways to be more sustainable. *The Water Treatment Magazine*. https://watertreatmentmagazine.com/en/impact-textile-industry/
8. Topics | European Parliament. (2020, December 29). *The impact of textile production and waste on the environment (infographics)*. https://www.europarl.europa.eu/topics/en/article/20201208STO93327/the-impact-of-textile-production-and-waste-on-the-environment-infographics
9. Topics | European Parliament. (2021, March 2). *How the EU wants to achieve a circular economy by 2050*. https://www.europarl.europa.eu/topics/en/article/20210128STO96607/how-the-eu-wants-to-achieve-a-circular-economy-by-2050
10. Uddin, F. (2019). *Introductory chapter: Textile manufacturing processes*. IntechOpen. https://doi.org/10.5772/intechopen.87968
11. American Chemical Society. (2012, October 26). *Global organic textile standard (GOTS)*. https://www.acs.org/education/whatischemistry/landmarks/rachel-carson-silent-spring.html
12. Intertek. (n.d.). *Global organic textile standard (GOTS)*. https://www.intertek.com/assurance/gots-textiles/
13. OTA. (n.d.). *Global organic textile standard (GOTS)*. https://ota.com/advocacy/organic-standards/fiber-and-textiles/global-organic-textile-standard-gots
14. Simple Ecology. (n.d.). *What is GOTS certified?* https://www.simpleecology.com/gots-certification
15. Ecocert. (n.d.). *Organic textile certification (GOTS)*. https://www.ecocert.com/en-IN/certification-detail/organic-and-ecological-textiles-gots
16. Blogger, P. P. (2024, April 15). *How fast fashion is bad for the environment*. Plastic Collective. https://www.plasticcollective.co/how-fast-fashion-is-bad-for-the-environment/
17. Maiti, R. (2024, May 30). *Fast fashion and its environmental impact in 2024*. Earth.Org. https://earth.org/fast-fashions-detrimental-effect-on-the-environment/
18. Nidhi, A. (2023, September 12). *Textile recycling: Techniques and challenges*. Fibre2Fashion. https://www.fibre2fashion.com/industry-article/9777/textile-recycling-techniques-and-challenges
19. Certilogo-Commerce. (2022, July 27). *Recycling textiles*. https://discover.certilogo.com/blogs/insights-blog/recycling-textiles-challenges-and-opportunities
20. Davis, J. (2024, August 28). *Our favourite sustainable brands to add to your wardrobe*. Harper's BAZAAR. https://www.harpersbazaar.com/uk/fashion/what-to-wear/g19491797/the-best-and-still-chic-sustainable-brands/
21. Statista. (2024, August 20). *Most sustainable luxury brands worldwide 2022*. https://www.statista.com/statistics/1323040/most-sustainable-luxury-brands/
22. Redress. (2019). *Zero-waste design technique*. https://redressdesignaward.com
23. Magdy. (2024, August 7). *Zero-waste fashion design*. IEREK. https://www.ierek.com/news/zero-waste-fashion-design/
24. Hallnås, L. (Ed.). (2019, November 29). *Zero waste design thinking*. https://www.diva-portal.org/smash/get/diva2:1316575/FULLTEXT02.pdf
25. Zero Waste Daniel. (n.d.). *About ZWD*. https://zerowastedaniel.com/pages/about#:~:text=ABOUT%20DANIEL,that%20send%20nothing%20to%20landfills
26. Good Fashion Fund. (2023, February 2). *Home*. Fashion for Good—Microsite. https://goodfashionfund.com/
27. Ellen MacArthur Foundation. (2024, May 22). *Press Release: Ellen MacArthur Foundation launches new initiative to make circular business models the norm*. https://www.ellenmacarthurfoundation.org/ellen-macarthur-foundation-launches-new-initiative-to-make-circular-business

Environmental Footprints of Luxury Textile Manufacturing

J. Jaisri and S. Balaji

1 Introduction

The core of luxury textile, an industry specifically associated with premium quality fiber and clothing for classy and luxurious people, has always been about sophistication, quality, and exclusiveness. Yet, with the rise in global consciousness of environmental sustainability, the industry has felt increasing pressure to confront its massive ecological footprint [1]. Producing luxury textiles, a sign of refinement used as a benchmark for high-end products, raises significant environmental issues due to its resource-intensive processes for superior materials. This is specifically reflected in the industry's carbon, energy, and water footprints [2]. The common thread linking these concerns is luxury textiles such as silk, wool, leather, and cashmere, all of which take enormous amounts of natural resources and energy to produce. The cultivation of these raw materials, along with the myriads of finishing techniques that determine the quality, finish, and texture of these fabrics, demands huge quantities of resources. For example, the growth of cotton and cattle breeding for wool or leather releases a considerable amount of carbon dioxide with relatively up their resource requirements in the form of freshwater and land [3]. In contrast, traditional dyeing and finishing stages are chemical and release a significant amount of water, polluting half with heavy metal substances [4].

The carbon footprint of a luxury textile is very evident in an industry that relies on processes that use a lot of energy and global supply chains. The energy needed to make fibers, process fabric, and transport it adds significantly to greenhouse gas

J. Jaisri (✉)
School of Creative Sciences, Costume Design and Fashion Department, Nehru Arts and Science College, Coimbatore, Tamil Nadu, India

S. Balaji
School of Commerce, Nehru Arts and Science College, Coimbatore, Tamil Nadu, India

Sadhna et al. (eds.), *Crafting Sustainability in Luxury Textiles for a Zero-Waste Future*, Sustainable Textiles: Production, Processing, Manufacturing & Chemistry, https://doi.org/10.1007/978-3-031-86773-6_2

emissions, especially fossil fuels being the main energy source. Moreover, since luxury textiles are typically sourced from different regions across the world with manufacturing often taking place across the globe (i.e., ship luxury textiles from the production country to the consumption country), the industry's carbon footprint is further exacerbated by the carbon footprint from long-distance transportation and logistics [5]. Another big problem is using water to make luxury textiles. Making raw materials, like cotton, uses a lot of water and relies on large amounts of fresh water for irrigation. Moreover, the processes of dyeing and finishing textile and garments use massive volumes of water and create severe water pollution [6, 7]. Because of these issues, more importance has been given to the study of the general impacts of luxury textiles through LCA. It is a very detailed process for looking at eco-friendly practices at every stage of the life cycle, from getting raw materials to making, using, and throwing away products. By breaking down eco-friendly aspects at each level of the supply chain, we can identify sectors using the most resources and find ways to reduce their impact. This approach is crucial in luxury textiles, where balancing quality and environmental responsibility is essential [5, 8].

The luxury textile industry also supports innovative environmental impact models that go beyond LCA methodology to further reduce its environmental footprint [9]. Measures like sustainable sourcing of raw materials, green manufacturing technologies, and circular economy practices reduce strain on the planet. These practices include using renewable energy, toxic-free materials and dyes, and closed-loop water systems [10]. Many luxury brands also support carbon offsetting projects and invest in renewable energy to offset manufacturing emissions. Driven by increasing consumer and corporate pressure for greater transparency, sustainability is now a priority for luxury textile manufacturers. However, the question remains whether these companies can redefine their environmental footprint. The luxury textile industry is in a pivotal position to reduce its impact by integrating life cycle assessment and environmental impact models into their processes while maintaining high quality for luxurious goods. This shift toward sustainable luxury is not only a response to consumer preferences but also essential for the long-term conservation of the industry and our planet [7, 11]. In this chapter, carbon footprint of luxury textile manufacturing will be analyzed with its parameters, sources, and impacts. Through examining processes of high-energy consumption, the sourcing of raw materials, transportation, and overall mitigation strategies, the environmental challenge associated with the industry is discussed in terms of its environmental impact, and the potential opportunities available for sustainable intervention.

2 Carbon Footprints in Luxury Textile Manufacturing

Carbon footprint is a metric used to quantify all greenhouse gas (GHG) emissions produced by humans. Limited additional relations remain similarly there that are used to quantify the emission of GHG but there the measurement system of representation varies slightly.

2.1 *Carbon Footprint and Parameters*

The carbon footprint is a measure of the total GHG produced from burning fossil fuel used for electricity, heat, cars, etc., from ton or kilogram or gram CO_2 equivalent. Global warming potential (GWP) is a comparative measure of how much high temperatures the GHG traps the environment, aids in the direction of global warming, and couples the degree of heat trapped by any relative mass of the gas being considered with the degree of heat deposited by rich mass of CO_2. This representation system can be helpful when we look at the state of nature throughout history. GWP is the impact of climate change over a certain time period, specifically in relative terms.

Duration (years)	Representation of GWP
20	GWP20
100	GWP100

Energy intensity is the ratio of overall energy consumption to gross domestic product (GDP).

2.2 *Carbon Footprint Types*

Figure 1 shows the types of carbon footprints. It is classified as primary carbon footprint and secondary carbon footprint. Primary carbon footprint results directly from the emission of greenhouse gases (GHG) during the combustion of fossil fuels such as coal, oil, and natural gas, and this category of carbon footprint is under our direct control. Primary carbon footprint is a result of carbon dioxide emissions from transportation, home energy use, and so on.

Secondary carbon footprint is another term for indirect GHG emissions created due to the life cycle of different products. It may occur in the use of clothing, reformation, and leisure things. The textile industry is identified as one of the largest producers of GHG all over the world [12]. It has been reported that textiles and aluminum produce the highest GHG emissions per unit of material. The textile sector is reported as the 6–8% highest emitter of CO_2 production. Luxury textile production, in particular, has been of great environmental issue because of this sector's giant carbon footprint illustrated by the leakage of greenhouse gases, particularly

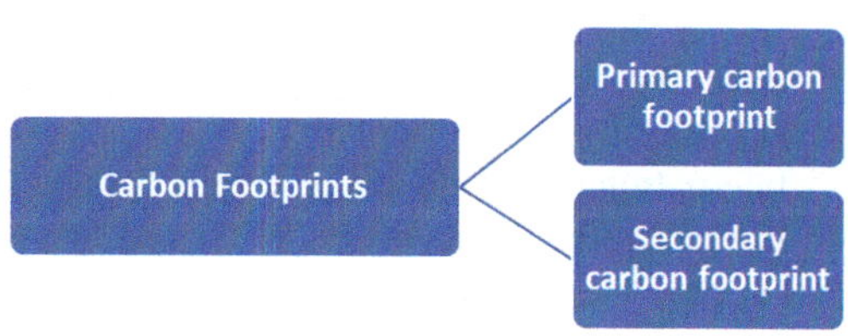

Fig. 1 Types of carbon footprints [12]

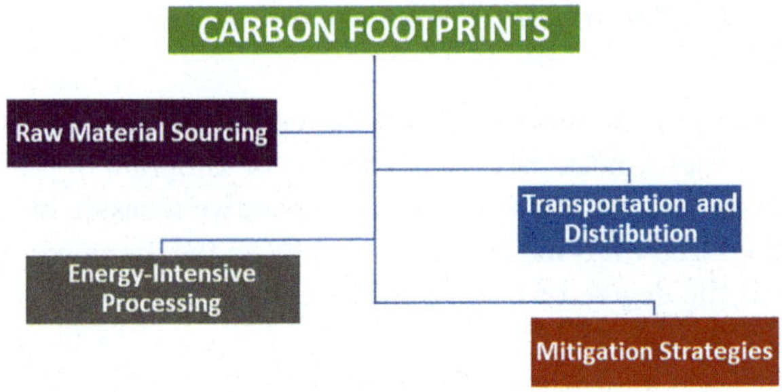

Fig. 2 Carbon footprints in luxury textile manufacturing [13, 14]

CO_2, when producing at a different stage in the production phase. These affect the global climates, and, in fact, the only textile sector noted for using precious resources [13, 14]. Figure 2 depicts various elements of carbon footprints, which comprises raw material sourcing, energy-intensive processing, transportation, and also the mitigation strategies for reducing environmental impact.

2.3 Raw Material Sourcing

The textile industry is primarily dependent on natural fibers for which the influence of environmental alteration is manifested on the development and production of these natural fibers. For the natural fibers, CO_2 is emitted throughout preparation, cultivation also field processes (weed control, mechanical irrigation, pest control, and fertilizers), harvesting, and yield. In the manufacture of natural fibers, commonly two different kinds of manures, such as fertilizer and artificial elements, are utilized. Synthetic manures lead to a big carbon footprint, making synthetic fertilizers one of the main components of conventional agriculture. However, 1 ton of nitrogen fertilizer has been known to cause the emission of about 7 tons of CO_2 equivalent GHG in its production. Table 1 shows CO_2 emissions of various textile fibers with their environmental impacts. Likewise, it delivers data on the emissions created by the methods involved [15].

So many organizations are designed to maintain the carbon footprint. Numerous administrations worldwide are dedicated to decreasing greenhouse gas (GHG) emissions and developing standardized methods for evaluating these emissions:

- California Climate Change Registry
- United Nations Framework Convention on Climate Change (UNFCCC)
- U.S. Department of Energy (DOE)
- Organisation for Economic Co-operation and Development (OECD)—International Energy
- World Resources Institute (WRI) and World Business Council for Sustainable Development (WBCSD)

Table 1 CO_2 emission details for various natural fibers and polyester fibers [15]

Fiber type	CO_2 emissions (kg CO_2 per kg of fiber)	Notes
Cotton	2–4	Includes agricultural inputs (e.g., water, pesticides, fertilizers)
Wool	7–10	Reflects methane emissions from sheep and processing energy
Linen (flax)	2–3	Lower emissions due to minimal fertilizer needs and energy-efficient processing
Jute	1–2	Low emissions due to rain-fed growth and limited energy-intensive steps
Silk	5–8	High emissions due to mulberry cultivation and energy-intensive reeling
Polyester	5–6	Fossil-fuel-based, energy-intensive manufacturing, but recyclable

- Intergovernmental Panel on Climate Change (IPCC)
- Lawrence Berkeley National Laboratory (LBNL)

2.4 *Intensive Energy Processing*

A major source of emissions in luxury textile manufacturing comes from processing as raw materials are turned into lush, luxurious fabrics through energy-intensive spinning, weaving, dyeing, and finishing processes. These processes require large amounts of energy, much of which is still generated from fossil fuels in many textile-producing regions. For example, silk fibers need to be boiled to remove them from the cocoon, a highly energy-intensive process, while wool requires scouring and heat treatment, both of which demand substantial [16, 17]. Dyeing, especially in high-end fabrics requiring bright and decorative color, also consumes high levels of energy due to the temperatures and chemicals involved [18–20]. Table 2 summarizes a few energy-intensive stages of some textile fibers and their impacts on the environment, which indicates potential areas for sustainable intervention.

2.5 *Transportation and Distribution of Carbon Footprint in the Textile Industry*

As shown in Fig. 3, luxury textiles with various textile fibers have their own carbon footprint related to transportation and distribution. Cotton and jute have less carbon footprints compared to synthetic fibers like polyester. Among all other textile fibers, silk plays a major role in carbon footprint on transportation and distribution. High-end fabrics and garments are often produced across different parts of the world,

Table 2 Energy-intensive processing in luxury textile manufacturing [18–20]

Material	Processing stages	Energy requirements	Environmental impact
Cotton	These are the steps: Ginning, spinning, weaving, dyeing, and finishing	Moderate (machinery energy use, water heating)	Abundant H_2O uptake, sewage manufacture, and die machinery dependency on fossil erudition
Wool	Scouring, heat treatment, spinning, weaving, dyeing, finishing	Thermal energy/water heating	CO_2 emissions, high-energy demand, and wastewater pollution caused by scouring agents Linen (flax)
Linen (flax)	Retting (soaking), scutching (fiber separation), spinning, weaving, dyeing, and finishing	Moderate (water and mechanical energy)	Water pollution during retting and moderate energy use for spinning and weaving
Jute	Retting, spinning, weaving, finishing	(Water consumption mechanical energy)	Retting effluent and CO_2 discharges from machinery use
Silk	Boiling cocoons, reeling, spinning, weaving, dyeing, and finishing	The processes: Boiling cocoons, reeling, spinning, weaving, dyeing, and finishing	Very high (thermal energy)

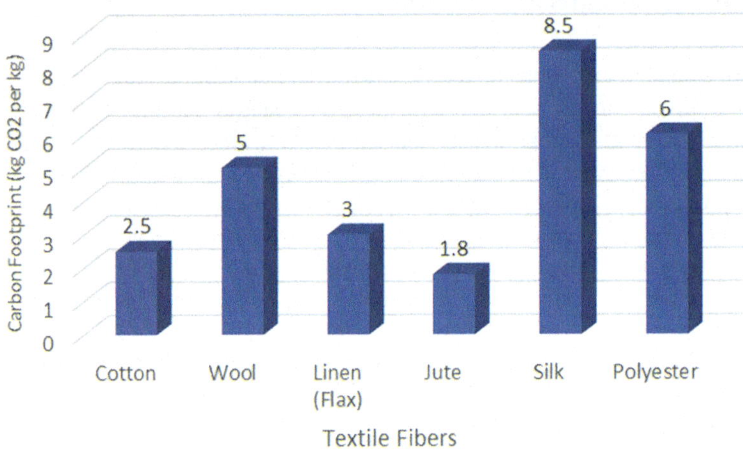

Fig. 3 Carbon footprints on transportation and distribution [21, 22]

which can promote sustainability but also means many materials may travel thousands of miles to reach their final destination. The luxury textile supply chain is complex, with raw materials grown or produced in one country, processed in another, and finally shipped to retail markets in a third. This extensive transport results in substantial carbon emissions as both air and sea transport are major emitters [23, 24]. Air transport is particularly carbon-intensive and is often chosen for high-end consumer goods due to the urgency associated with luxury products [21, 22]. For reducing environmental impact, it is better to choose low-carbon impact fibers like jute and linen.

2.6 Mitigation Strategies

Humanity has the collective responsibility of reducing the carbon footprint and creating a sustainable environment for future generations. As the civilizations progressed, the textile went from being just a basic need to becoming an expression of lifestyle and fashion, leading to increased per capita clothing consumption. This trend, fueled by improved living standards and growing consciousness about fashion, has magnified the environmental footprint of the textile industry. It involves all the life cycle stages, from raw material production to waste disposal, for a textile product. The world's largest industrial sector and major source of greenhouse gas emissions are the textile industries [25].

Efforts to mitigate the eco-friendly effects include adopting vigor well-organized and water-saving technologies. Luxury textile producers are increasingly employing innovative processes and intermediate products that reduce energy usage. These advancements span the entire supply chain, from harvesting and processing to packaging, transportation, and disposal. The production of synthetic fibers, although widely used, is energy-intensive, making the use of natural fibers an essential corrective measure. By prioritizing natural fibers and integrating sustainable practices at every stage, the industry can significantly lower its carbon footprint while maintaining product quality and consumer appeal [26].

2.6.1 Alternate Fabrics

Unlike synthetic fibers, the natural fibers are very low in carbon footprint. Organic fibers are used that reduce the carbon footprint. Making a lower carbon footprint is a process that takes less energy. They are also bestowed with other priceless benefits such as biodegradability and carbon sequestration from the atmosphere. Cotton uses more pesticides than any other US-grown crop (and polyester uses 70,000,000 barrels of oil a year). Crops such as flax and hemp, as well as even "spider silk," are being viewed more and more as an alternative to traditional fabrics as they are now much more sustainable. These textiles are primarily owing to its characteristics such as conductivity and high-tensile strength that suit them for use in technical applications [27, 28].

2.6.2 Production of Technical Textiles with Resources

The production and processing of traditional textiles require significant resources, including energy and water, contributing to a high carbon footprint. However, synthetic textiles like acrylic fiber, which can be produced in-house with standard chemical compounds and specialized equipment, offer an alternative. These fibers can be manufactured to close tolerances, making them suitable for bespoke or small-scale production. Textile processing remains a major contributor to the carbon footprint due to its resource-intensive nature. Strategies to mitigate this impact include adopting water- and energy-efficient technologies. Low and ultralow liquor ratio machines save water throughout the pretreatment, dyeing, washing, and finishing operations while using lower amounts of heating energy and with reduced effluent treatment loads [29].

Resource conservation also occurs due to advancements in dyeing and finishing technology, continuous dyeing machines for knitwear, among others. Eco-efficient chemicals and solutions help reduce the processing steps that contain dyeing, washing, and finishing, thereby requiring lesser amounts of energy and water. New processes like singeing–desizing in bath dyeing for polyester/cotton blends help reduce the steps required in the processing of textiles, thereby saving energy and water. Other methods include thermosol dyeing and cold pad batch (CPB), which have improved efficiency in using resources. Enzyme-based processing techniques have also replaced harsh chemicals used in these production lines. The result of this production is decreased energy consumption, water usage, and environmental impact. High-fixation reactive dyes and digital inkjet printing technologies also facilitate the industry's shift to lower carbon footprints, showing a decline in demand for water and chemicals [30].

2.6.3 Synthetic Materials "Upcycling"

Upcycling is revolutionizing recycling by finding innovative applications for old materials through advanced extrusion technologies. For instance, nontextile products such as beer bottles and polypropylene bags can now be processed into fine particulates, melted, and extruded into usable fibers. This approach not only minimizes waste but also reduces the transportation distances for sourcing materials, thereby lowering carbon emissions. Technological advancements have been paired with strategies like textile reuse and recycling to significantly reduce clothing industry carbon footprint. However, recycling is only sustainable if the carbon emissions from the recycling process are lower than emissions from disposal. Methods such as donating clothes for secondhand use in global markets contribute to reducing waste while extending product life cycles. Analyzing energy use and greenhouse gas (GHG) emissions from collection, sorting, and selling of used textiles compared to producing new items is crucial for validating the environmental benefits of recycling [30].

Furthermore, carbon footprint reduction in the textile sector can be achieved through recycling process waste and end-of-life textile products. Incorporating organic natural fibers, which typically have a lower carbon footprint than synthetic fibers, is gaining traction as a sustainable approach. While the cultivation of natural fibers emits CO_2 and requires significant energy, the synthetic fiber production phase has higher energy consumption and CO_2 emissions. Transporting raw materials and finished products also adds to the industry's carbon footprint, but solutions like composite textile mills and organic farming hold promise for future sustainability.

3 Life Cycle Assessment (LCA)

LCA is a durable device for ecofriendly impact analysis and can measure these emissions and energy use throughout the life cycle of a product. Advanced LCA software helps in understanding and minimizing the environmental impact of the textile sector, which is critical because this energy-intensive industry seeks greener solutions [26].

It measures the latent effect by the

- Relevant inputs and outputs
- Evaluation of the environmental impacts regarding the inputs and outputs
- Outcome of the inventory and evaluation interpretation

3.1 Life Cycle Assessment Stages

As shown in Fig. 4, the stages of a life cycle assessment (LCA) are critical for understanding the eco-friendly impression of products and services.

- **Goal and scope**: The objectives and boundaries of the assessment are defined.
- **Life cycle inventory (LCI)**: Collection of data occurs in this phase, documenting the materials, energy flows, and environmental interactions within the product's life cycle. It identifies raw material usage and emissions.
- **Life cycle impact assessment (LCIA)**: This stage evaluates the ecological things of the information collected in the list, including emission measurements and their potential ecosystem and human health impacts, using methods like standardization and weighting.
- **Interpretation**: It involves summarizing the impression valuation and suggesting strategies to mitigate environmental effects [15].
- To ensure accuracy, this process adheres to guidelines from the International Organization for Standardization [31].

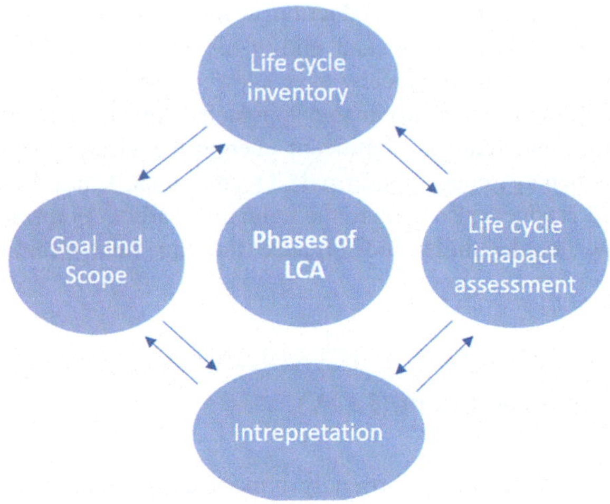

Fig. 4 Stages of a life cycle assessment [15, 31]

3.2 Life Cycle Assessment Uses

Life cycle assessment would determine the sustainable effects and can point out some opportunities for the reduction of impacts. LCA would contribute to the new product development as well as study the environmental impact of many different materials and manufacturing processes. It could be used in the management of a supply chain, along with procurement efforts that focus on seeking more sustainable suppliers or products. Also, determine the environmental performance of your suppliers and publicly declare the opportunities for supply chain reductions in environmental impacts. For an application domain, use LCA data in marketing and sales to deliver the environmental product approval, such as EPA's Energy Star Certification. It enables one to demonstrate the ecological value of a product in a manner that makes LCA data available for executives and strategic managers to make decisions about pollution prevention.

4 Conclusion

The luxury textile industry faces a make-or-break moment as it attempts to balance its age-old focus on exclusivity and quality with heightened demand for environmental respect. This chapter highlights some of the immense environmental issues brought about by the sector, such as the creation of a huge carbon footprint, high water usage, and the sector's dependence on fossil fuels. We can make the industry more sustainable by systematically analyzing and possibly improving the ecological impact of the industry at every stage of its life cycle—from sourcing natural resources, processing, to distribution—through life cycle assessment. Implementing sustainable materials, energy-efficient technologies, and circular economy principles are keys to combating these hurdles all of which contribute to sustainable construction. Technological advancements in upcycling, dyeing with less water, and powering factories with renewable energy show how we could still truly lower the industry's environmental impact without sacrificing product quality. The demand is forcing an evolution within the high-end fashion landscape, but the shift is becoming critical for survival and responsible break as an industry. Implementing these measures would allow the luxury textile industry to establish a new position in the global landscape and become a benchmark for responsible production and consumption.

References

1. Smith, J., et al. (2021). The environmental sustainability challenges of luxury textile production. *Journal of Sustainable Luxury Practices, 29*(3), 112–127.
2. Jones, R., & Chen, L. (2020). Carbon, energy, and water footprints in high-end textile manufacturing. *Environmental Textiles Review, 18*(2), 89–103.
3. Brown, W. W. (2019). Resource demands and ecological impacts of luxury textile production. *Journal of Environmental Textiles, 18*(3), 102–119.
4. Green, T., & Lee, J. (2021). The carbon footprint of luxury textiles: A global supply chain perspective. *Environmental Textiles and Practices, 20*(2), 87–101.
5. Kim, H., & Green, T. (2021). LCA and beyond: Transforming sustainability in luxury textile manufacturing. *Journal of Sustainable Practices, 29*(3), 67–82.
6. Jones, R., et al. (2021). Water-intensive processes in luxury textiles: Challenges and solutions. *Journal of Environmental Textiles, 23*(2), 144–160.
7. Zhao, Q., & Wei, L. (2022). Global transportation and the carbon footprint of luxury textiles. *International Journal of Environmental Studies, 27*(4), 198–214.
8. Zhao, Q., & Lee, P. (2022). Integrating life cycle assessment into luxury textile practices. *Journal of Sustainable Innovations, 31*(5), 123–140.
9. Smith, J., Brown, A., & Johnson, L. (2022). Environmental impacts of luxury textile manufacturing: Life cycle analysis and sustainability challenges. *Journal of Sustainable Practices, 34*(2), 123–145.
10. Brown, W., & Johnson, T. (2020). Sustainable practices in the luxury textile industry: Toxic-free dyes and renewable systems. *Journal of Sustainable Innovations, 22*(4), 213–230.
11. Green, T., et al. (2020). Circular economy in luxury textiles: Towards sustainable sourcing and closed-loop systems. *Journal of Textile Sustainability, 15*(1), 55–72.
12. Kissinger, M., et al. (n.d.). High GHG emissions in textiles and aluminum: A comparative study. *Environmental Impact Studies, 12*(3), 45–60.
13. Brown, W., & Johnson, T. (2021). Carbon footprints and the environmental challenges of the luxury textile industry. *Journal of Environmental Textiles, 19*(4), 217–232.
14. Smith, J., et al. (2022). Greenhouse gas emissions in luxury textile production: Impacts and mitigation strategies. *Journal of Sustainable Luxury Practices, 29*(5), 123–140.
15. International Organization for Standardization. (2006). *ISO 14040: Environmental management—Life cycle assessment—Principles and framework*. ISO.
16. Brown, W., & Smith, J. (2021). Energy-intensive processes in luxury textiles: Challenges and solutions. *Journal of Textile Sustainability, 25*(3), 89–105.

17. Kim, H., Lee, P., & Zhao, Q. (2019). Environmental impact of silk and wool processing: Energy requirements and mitigation strategies. *Journal of Environmental Textiles, 21*(2), 78–92.
18. Jones, R., et al. (2020). Energy demands and environmental impacts of dyeing in high-end textiles. *Journal of Sustainable Textile Practices, 18*(4), 56–72.
19. Lee, P., & Chen, L. (2018). Advanced dyeing technologies and their environmental implications. *International Journal of Textile Science, 14*(1), 123–140.
20. Zhao, Q., & Wei, L. (2021). Sustainable approaches to energy use in textile processing. *Journal of Green Manufacturing, 16*(3), 45–63.
21. Kim, H., & Chen, L. (2022). Environmental challenges in global luxury textile supply chains. *Journal of Textile Sustainability, 22*(3), 134–148.
22. Smith, J., et al. (2021). Carbon footprint mitigation strategies for luxury textiles. *Journal of Sustainable Luxury Practices, 20*(4), 112–130.
23. Brown, W., & Lee, P. (2019). Carbon emissions in luxury textile transportation and distribution. *Journal of Sustainable Fashion, 15*(4), 89–104.
24. Zhao, Q., & Wei, L. (2021). Energy-intensive processes in luxury textile production: A global review. *Journal of Green Manufacturing, 16*(3), 67–83.
25. Smith, J., & Kumar, A. (2023). The environmental impacts of per capita clothing consumption. *Journal of Environmental Sustainability, 28*(5), 90–105.
26. UNEP (United Nations Environment Programme). (2023). *Sustainable practices in the global textile industry: Reports and guidelines*. UNEP.
27. Bier, M. (2021). Sustainable fabrics: The rise of natural and alternative fibers in the textile industry. *Green Textiles Journal, 12*(3), 45–57.
28. Shen, L., & Wang, Z. (2020). Environmental impacts of textile fibers: Natural vs synthetic fibers. *Journal of Sustainable Fashion, 5*(2), 32–45.
29. Gupta, R., & Sharma, M. (2023). Eco-efficient technologies for sustainable textile processing. *Journal of Environmental Textiles, 29*(2), 145–162.
30. Jain, S., et al. (2023). Innovations in dyeing and finishing technologies: Environmental implications. *Journal of Green Manufacturing, 17*(1), 56–71.
31. International Organization for Standardization. (2006). *ISO 14044: Environmental management—Life cycle assessment—Requirements and guidelines*. ISO.

Sustainable Practices in Clothing Supply Chain

S. Greeshma, Sadhna, and Rajesh Kumar

1 Introduction

Sustainable practices in the clothing supply chain mean that during the entire life cycle of fashion products, from raw material sourcing to production, distribution, and disposal, companies must integrate environmentally friendly and socially responsible methods. It focuses on practice reducing adverse environmental and social impacts and ensuring fair wages for workers. Sustainability, as defined, involves "meeting the needs of the present without compromising the ability of future generations to meet their own needs," encompassing environmental, social, governance, and economic aspects [1].

Sustainability is a key feature in the fashion industry, which is known for the immense environmental impact, such as pollution, resource depletion, and unethical labor practices. The latest research indicates that 62% of Gen Z consumers prefer brands that value sustainability, and 73% are willing to pay more for eco-friendly products [2]. This shift in consumer behavior reflects a growing awareness of the ethical implications of fast fashion and a demand for more responsible production practices. As such, brands are increasingly reevaluating their supply chains to adopt sustainable practices that align with consumer expectations and regulatory pressures.

Material selection is the first step toward minimizing the environmental impact of the fashion industry as raw material production accounts for 38% of total emissions. Organic cotton and recycled polyester come with minimal energy expenditure and pollution. The recycled polyester product is developed using post-consumer plastic waste that discourages waste from lodging in landfills and reduces reliance on virgin resources. Initiatives such as the Material Change Insights report by

S. Greeshma (✉) · Sadhna · R. Kumar
Department of Fashion Design, School of Arts and Design, Woxsen University, Hyderabad, Telangana, India

Sadhna et al. (eds.), *Crafting Sustainability in Luxury Textiles for a Zero-Waste Future*, Sustainable Textiles: Production, Processing, Manufacturing & Chemistry, https://doi.org/10.1007/978-3-031-86773-6_3

Textile Exchange show growth; cotton sourced from organic or recycled sources was at 66.16% in 2022. Brands can not only minimize their carbon footprint but also tap into a rapidly growing market of environmentally conscious consumers through the use of eco-friendly materials [3].

Sustainable clothing supply chains and efficient production processes are essential in reducing waste and conserving energy. Traditional manufacturing processes can use much water and produce chemical waste, but new approaches aim to minimize these impacts. For example, companies like Levi's incorporated water-saving technologies into their denim manufacturing processes [1].

Transparency, secondly, will assist the brand in maintaining labor practices and environmental responsibility at each stage of the supply chain. Clear visibility from sourcing raw materials to retail facilitates effective management of labor and environmental risks. Open-source predictability protects the reputation of the brand and allows high consumer trust [4].

The circular economy is the very basis of achieving sustainability in fashion. This model encourages designing products for long life, reuse, and recyclability rather than a linear "take-make-dispose" approach. More and more brands seek circularity as part of their business models, introducing take-back schemes or creating items that can be quickly taken apart for recycling. Circular economy principles not only help decrease waste, but also trigger a shift in the quantity-to-quality consumption culture. Brand mindfulness demands systematic ownership of the product life cycle [5]. Fostering a culture of mindful consumption can thus facilitate the sector's shift toward a durable and resource-optimal sustainable system.

2 Understanding Sustainability in Fashion

Fashion sustainability is aimed at minimizing the negative impacts of the production and consumption of clothes on the environment and society, which are driven by essential concepts promoting ethical and eco-friendly behavior. Supply chain transparency is necessary to hold brands accountable for the sourcing of all materials and production methods; ethical labor demands all workers are paid fairly and in safe conditions; and collaborative consumption reduces waste by implementing the sharing, renting, or swapping of clothing. Circularity is about creating clothes that are designed to last, repairable, and recyclable; thus, eco-friendly materials such as organic cotton and recycled polyester prevent harmful impacts on the environment [6]. Figure 1 depicts the circularity in fashion. Collectively, these principles support a fashion sector that is good for people and the planet.

Traditionally, the fashion industry operates on a linear "take-make-dispose" approach, contributing to resource depletion and overconsumption of pollution by using products briefly before discarding them. Circular business models (CBMs) prompt a rethink of the above by reconceptualizing the product life cycle, emphasizing durability, repair, reuse, and recycling. CBMs focus on reducing their environmental footprint and developing creative ways to keep materials in the production sphere longer.

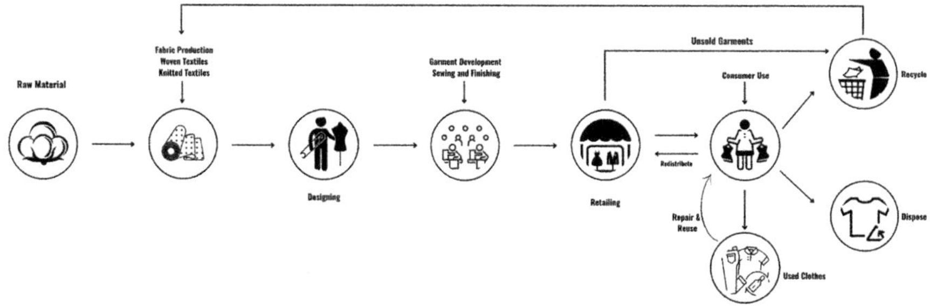

Fig. 1 Circularity in fashion [7, 8]

In the context of circularity in the fashion sector, product life extension and recycling initiatives must focus on challenges to their necessary evolvement. Strategies such as garment leasing, repair, and upcycling are examples of product life extension that minimizes waste and maximizes the value of the resources used. Recycling efforts look to reclaim materials for reuse, such as when used textile waste is made back into reusable fibers or used to create innovative new materials, like at Renewcell and Vegea [9]. However, the transition faces challenges, including limited recycling technology, textile-to-textile recycling for complex blends, limited and inadequate collection and sorting infrastructure, and a need for more consumer awareness. Another has been the cost and perception of recycled material as inferior.

Consumer behavior has greatly influenced sustainability trends in fashion; rising environmental awareness will force demand for sustainable options.

By 2030, it is estimated that 50% of consumers will prioritize sustainability, a significant increase from the current 15% [10]. Still, the nitty-gritty of price, quality, brand reputation, and style tends to overwhelm environmental issues. Younger generations, such as Millennials and Gen Z, are more environmentally conscious but also have budgets to manage, so it is up to brands to balance affordability with eco-friendliness. There is an "attitude–behavior gap" where concern for sustainability does not consistently translate into purchasing behavior, mainly because of convenience, product availability, and social pressure. To overcome these issues, brands must be transparent in their practices, affordability should not compromise sustainability, and marketing strategies must effectively target conscious consumers. Therefore, by aligning with consumer preferences, fashion companies can meet the growing demand for sustainable products while driving meaningful change within the industry [11].

3 Innovations in Sustainable Apparel Production

Eco-friendly materials are revolutionizing sustainable apparel production by cutting the fashion industry's environmental footprint through organic, recycled, and biodegradable fabrics. Organic materials include cotton, grown without synthetic pesticides or fertilizers, and hemp, which requires fewer resources and improves soil quality. Recycled materials include recycled cotton and ECONYL® recycled nylon, which reduce landfill waste, save resources, and minimize the environmental impact by utilizing post-consumer waste. Innovations in biodegradable material, such as Pinatex from pineapple leaf fibers, banana-tex from banana plant fibers and apple leather, made from the by-product of apple juice, present sustainable, animal-free options that decompose naturally while utilizing agricultural waste. These materials reduce resource usage and minimize greenhouse gas emissions and chemical consumption associated with traditional textile manufacturing [12]. The fashion industry moves toward a more sustainable ecosystem by embracing such innovations, showing how thoughtful material choices can significantly lessen its environmental footprint and promote responsible production practices.

Fashion technology is revolutionizing the apparel industry by transforming design, production, and consumer engagement through CAD, CAM, and AI innovations. Designers leverage these tools to craft unique, precise patterns efficiently, while advancements like 3D printing enable custom-fit garments, complex geometries, and reduced waste. Wearable technologies, innovative fabrics, and embedded electronics blend aesthetics with functionality, adding new dimensions to fashion. Sustainability is at the center as brands such as Zara and Adidas focus on going green by employing biodegradable textiles, using recycled materials, and developing nanotechnology. AI enhances inventory management while predicting trends and improving customer shopping experience through chatbots, virtual try-ons, and live streaming. Digital clothing, depicted through avatars on the likes of Roblox and The Fabricant, reveals the harmony between fashion and technology [13]. Emerging trends, such as lab-grown leather, IoT-integrated apparel, and AR-enabled shopping, are changing the industry. Merging innovation with sustainability, fashion technology creates a dynamic, accessible, and eco-conscious future for global consumers.

Fashion, often associated with glamour and trends, is changing its focus today, focusing on sustainability and ethics as key issues for the future. Ethical and sustainable clothing is all about sourcing and manufacturing clothes that benefit society and minimize destructive environmental impact. The bottom line begins with organic growth and avoids pesticide use through conventional cotton farming, whereas processes should be transparent concerning fair wages, worker security, and environmental protection [14].

Pioneering brands like Stella McCartney show how style goes hand in hand with sustainability, and organizations like Ethical Fashion Forum raise worldwide awareness and set the standards for the ethical approach. However, "greenwashing" causes consumers to lose trust. Therefore, it is necessary to have transparency in this area. Ethical fashion supports workers, reduces carbon footprint, and avoids animal cruelty. Consumers can play the most significant role by wearing durable, eco-friendly materials and adopting cold-water washing and air-drying techniques [15]. With the leading brands embracing sustainable methods, the future of ethical fashion promises to become an industry standard in favor of society and the planet [16]. Organizations such as the Ethical Fashion Forum (EFF) play a vital role in the industry for its sustainable practice. EFF has been highly instrumental in raising awareness regarding ethical fashion, providing resources to brands adopting sustainable methods, and promoting excellence through awards [17]. Still, challenges continue to arise, most especially concerning "greenwashing," exaggerating the efforts of the brand toward sustainability in the hopes of deceiving customers. This calls for distrust and highlights the need for transparency in the fashion world [18].

Ethical fashion supports workers' rights, reduces carbon footprints, and prevents animal cruelty. Consumers can contribute significantly to this movement by opting for durable, eco-friendly clothing options and practicing cold-water washing and air-drying. With more leading brands committing to sustainable methods, ethical fashion will become the standard in the industry and benefit society and the planet [19].

4 Enhancing Energy and Water Efficiency

Fast fashion is characterized by low-cost, mass-produced clothing that has a significant impact on the environment. It takes 2700 L of water to make a single cotton t-shirt that can quench one person's thirst for 2.5 years. The EU textile industry consumed 9 m^3 of water, 400 m^2 of land, and 391 kg of raw materials per citizen in 2020. Textile production also contributes to 20% of global clean water pollution through dyeing and finishing processes. Synthetic fabric washing releases around 700,000 microplastic fibers per load, contributing over 500,000 tons of microplastics into the oceans annually. Fast fashion also produces significant emissions of greenhouse gases: the EU purchases of textile materials produced 270 kg of CO_2 per person in 2020. Europeans throw away 11 kg of textiles yearly, while 87% are burnt or sent to landfills, but only 1% are recycled into new garments [20].

The Indian textile industry is growing at a CAGR of over 15% to achieve $100 billion in exports by 2030 and create 34 million jobs. It employs 4.6 crore people and faces challenges such as generating 5% of global landfill waste and being the second-largest polluter after oil. India accounts for 25% of the global cotton-growing areas but produces only 8% of cotton, and the country is highly dependent on expensive imports. The weaving sector is largely unorganized, with over 90% outdated infrastructure and minimal automation. Skill shortages, rising raw material costs, and environmental noncompliance further hinder growth, which requires collaborative efforts to overcome these issues [21].

The global textile industry, valued at $1.83 trillion in 2023, must adopt sustainable solutions to reduce its environmental footprint. Some key measures include transitioning toward circular business models emphasizing reuse, recycling, and material recovery. Companies should also focus on extending garment life through rental, secondhand resale, and redesign services. Investments in advanced material recovery technologies and adopting life cycle assessment would guide choices in more ecologically friendly fabrics, such as recyclable polyester, rather than water-intensive cotton. Responsible sourcing, transparent supply chains, and adherence to ethical standards such as SA 8000 will also play a critical role. Recovery of water with an efficiency of 95–98% and biological treatment systems will offset water pollution. Microplastics reduction practices through innovative textile washing practices are in demand [22]. The adoption of durable fabrics, waste reduction, and building a responsible consumer culture will be other steps toward the transition of the industry. Sustainable innovation in the textile sector, therefore, can lead to sustainable solutions to environmental challenges while fulfilling the responsibilities in growing global demand.

5 Circular Economy in Fashion

Investments in advanced material recovery technologies and life cycle assessment will be used to determine which of the more sustainable fabrics to use instead of water-intensive cotton, such as recyclable polyester. Responsible sourcing, transparent supply chains, and compliance with ethical standards like SA 8000 are a must.

Technologies that recover water with 95–98% efficiency and biological treatment systems will address water pollution. In addition, innovative practices for reducing microplastics in textile washing are required. Embracing durable fabrics, reducing waste, and creating a responsible consumer culture will be further steps toward the industry's transition. In this way, the textile sector can lead sustainable innovation, addressing environmental challenges while meeting growing global demand responsibly [23].

Circularly, design is the very first factor that promotes sustainability. Designers are challenged to create durable garments resistant to trends and time so that they may be replaced less frequently. This makes consumers value quality rather than quantity. Moreover, consumers are increasingly shifting toward resale and rental models. The clothing rental services allow one to enjoy variety without actually owning it. This aligns with the desire for experiences rather than possessions, a growing phenomenon. This phenomenon is mainly seen among the younger generation, where sustainability and flexibility in choice are the highest priorities for purchase. By educating consumers on how to care for their clothes through repair services and maintenance, for example, the life cycle of products can be extended further. This move will reduce waste and create a conscious consumption culture where consumers engage in sustainable practices [24].

The circular economy has inspired innovative business models in fashion, where sustainability and economic benefits coincide. These include clothing subscription services and traceable garment initiatives. Subscription platforms enable consumers to rent garments, which means variety and affordability but also reduces waste and the environmental impact of fast fashion. Traceable garment initiatives supported through Digital Product Passports would offer transparency at all life cycle stages for a garment, allowing consumers to make more informed purchase decisions and pushing brands toward responsible sourcing. This approach will not only solve environmental problems but also unlock market spaces focused on sustainability, where customers can become loyal [25]. By embracing such an approach, brands position themselves at the forefront of responsible consumption and show commitment toward the environment and innovative growth in business.

6 Conclusion

Transitioning toward the sustainability of the clothing supply chain is no longer an option but an imperative for the fashion world. Increasing consumer awareness of the purchases' environmental and social implications requires brand adjustment in response to the emerging consumer. Moving from the traditional linear model to the circular economy involves designing products, producing, and consuming in a completely different way. This approach has longevity, repairability, and recyclability principles, which can significantly minimize waste and resource depletion. Sustainable practices include using environmentally friendly materials, such as organic cotton and recycled fabrics. This reduces environmental impact and caters to the growing market of conscious consumers who prefer sustainable purchase choices. Technological innovations are vital in furthering sustainable apparel production, improving efficiency, and reducing waste through 3D printing and AI-driven inventory management.

However, there are problems. Examples of "greenwashing" are brands exaggerating sustainability practices to mislead the consumer, undermine trust, and demand transparency in the industry, among other things. Brands will have to prioritize ethical practices that ensure fair wages for all workers involved in a supply chain and safe working conditions. It will take cooperation between manufacturers, retailers, consumers, and policymakers to change the industry. Partnerships must be encouraged to help bring sustainability practices and more transparency into the supply chain.

References

1. Purohit, R. (2023, August 11). *Towards sustainable fashion apparel supply chains*. Observer Research Foundation. https://www.orfonline.org/expert-speak/towards-sustainable-fashion-apparel-supply-chains
2. Savilia, K. (2024, April 26). *The importance of sustainability in fashion*. Forbes Technology Council. https://www.forbes.com/councils/forbestechcouncil/2024/04/26/the-importance-of-sustainability-in-fashion/
3. Poratelli, D. F. (2023, August 2). *Sustainable materials in the fashion industry: New data from textile exchange*. CIKIS STUDIO. https://www.cikis.studio/en/article/sustainable-materials-in-the-fashion-industry-new-data-from-textile-exchange
4. Vicinus, A. (2022, November 15). *How sustainability affects the fashion supply chain—State of matter apparel*. State of matter Apparel. https://stateofmatterapparel.com/blogs/som-blog/how-sustainability-affects-the-fashion-supply-chain

5. Kaur, R. (2023). Impact of circular economy on sustainable fashion: Opportunities and challenges. *International Journal for Research in Applied Science and Engineering Technology, 11*(9), 59–63. https://doi.org/10.22214/ijraset.2023.55602
6. Bardon, A. (2024, May 31). *Sustainable fashion: Eco-friendly practices in the industry | LU*. Lindenwood University Online. https://online.lindenwood.edu/blog/sustainable-fashion-exploring-eco-friendly-practices-in-the-industry/
7. Lane, H. (2022, May 12). *Circularity in fashion—Redress design award*. Redress Design Award. https://www.redressdesignaward.com/academy/resources/guide/circularity-in-fashion
8. Jackalo. (n.d.). *Circularity*. Jackalo. https://hellojackalo.com/pages/circularity
9. D'Itria, E., & Aus, R. (2023). Circular fashion: Evolving practices in a changing industry. *Sustainability: Science, Practice and Policy, 1*, 2220592. https://doi.org/10.1080/15487733.2023.2220592
10. Noushad, S. (2023, August 27). *Influencing change: How consumer behavior drives sustainable fashion market trend*. Entropik. https://www.entropik.io/blogs/fashion-market-research
11. Ankit Guria, A., & Roopa, K. V. (2024). Research paper on sustainable fashion and consumer buying behavior. *International Journal of Research and Analytical Reviews (IJRAR), 11*(1), 531–533.
12. Gusain, S. (2024, June 4). *The revolution of sustainable fabrics in fashion*. Beyours. https://www.beyours.in/blogs/news/the-revolution-of-sustainable-fabrics-in-fashion
13. Soliwal, T. (2024, January 9). *Fashion tech: How technology is shaping the future of apparel*. Fibre2Fashion. https://www.fibre2fashion.com/industry-article/9886/fashion-tech-how-technology-is-shaping-the-future-of-apparel
14. Sharma, R. (2024, September 4). *Why sustainable fashion is crucial for a greener future and ethical consumerism*. Gofynd the Closet. https://www.gofynd.com/thecloset/why-is-sustainable-fashion-important
15. Bennett, E. (2024, July 18). *The difference between sustainable and ethical fashion and why we need both*. Econyl. https://econyl.aquafil.com/eco-design-magazine/the-difference-between-sustainable-and-ethical-fashion-and-why-we-need-both/
16. Lolli, A. (2023, September 6). *Stella McCartney: Fashion and sustainability*. FG Conscious Fashion. https://thefashionglobe.com/stella-mccartney/
17. *Projects—Ethical Fashion Forum*. (n.d.). Ethical Fashion Forum. https://the.ethicalfashionforum.com/projects
18. Singh, S. (2024, June 27). *Greenwashing in the fashion industry*. Carbon Trail. https://carbontrail.net/blog/greenwashing-in-the-fashion-industry/
19. Li, M., Choe, Y., & Gu, C. (2024). How perceived sustainability influences consumers' clothing preferences. *Scientific Reports, 14*(1), 28672. https://doi.org/10.1038/s41598-024-80279-4
20. European Parliament. (2022, March 21). *The impact of textile production and waste on the environment (infographics)*. European Parliament. https://www.europarl.europa.eu/topics/en/article/20201208STO93327/the-impact-of-textile-production-and-waste-on-the-environment-infographics
21. Fogla, H. (2023, June 2). *Top 6 challenges faced by the textile industry in India*. Jumac Cans. https://www.jumaccans.com/blog/top-6-challenges-faced-by-the-textile-industry-in-india/
22. Prehu, I. (2024, April 2). 5 environmental impacts of textile industry, and 3 ways to be more sustainable. *The Water Treatment Magazine*. https://watertreatmentmagazine.com/en/impact-textile-industry/
23. Recovo. (n.d.). *How to implement the circular economy in the textile industry*. Recovo. https://recovo.co/en/blog/article/how-to-implement-the-circular-economy-in-the-textile-industry
24. Ilic, P. (2023, June 19). *How will circular fashion create new business opportunities*. PSQR. https://psqr.eu/publications-resources/circular-fashion/
25. Sustainable Fashion Forum. (2023, December 10). *What is circular fashion?* https://www.thesustainablefashionforum.com/pages/what-is-circular-fashion

Functional Longevity and Reusability of Luxury Textile Products

J. Jaisri and S. Greeshma

1 Introduction

Functional longevity in the luxury fashion realm reflects a new dimension of sustainability that fuses the physical durability of fabrics with their prevalence in the minds of consumers [1]. This duality challenges the current paradigms of fast fashion, highlighting clothing as long-lasting artifacts rather than temporary items. Functional longevity not only tackles looming issues of textile waste and resource depletion but encourages a more substantive emotional connection between consumers and their purchases [2]. With growing conversations around sustainability, luxury fashion rises as an industry both responsible and positioned to advocate for longer garment lifetimes via better materials, superior craftsmanship, and a more timeless aesthetic. The sustainability benefits of functional longevity are persuasive. Wearing a garment twice as long, on average, can cut its ecological footprint by 50% [3]. To be a luxury means to be a resource, at least, that should be a luxury to be available and the opposite of it, the fast fashion is just so wasteful [4]. Aside from environmental benefits, the concept also reinforces brand equity and consumer loyalty as long-lasting garments become physical manifestations of quality, exclusivity, and individual importance [5]. Utilizing tangible longevity in the design and production process for luxury brands requires a mature synthesis of pioneering materials, precision mechanics, and timeless proportions [6].

J. Jaisri
School of Creative Sciences, Costume Design and Fashion Department, Nehru Arts and Science College, Coimbatore, Tamil Nadu, India

S. Greeshma (✉)
Department of Fashion Design, School of Arts and Design, Woxsen University, Hyderabad, Telangana, India

Sadhna et al. (eds.), *Crafting Sustainability in Luxury Textiles for a Zero-Waste Future*, Sustainable Textiles: Production, Processing, Manufacturing & Chemistry, https://doi.org/10.1007/978-3-031-86773-6_4

The global luxury textile market size was valued at around US$80 crores in 2023 and is estimated to grow to over US$151.32 crores by 2029 at a CAGR of 4.1% from 2024 to 2030. Several factors ranging from increased disposable incomes, rapid changes in consumer tastes and preferences, and impact of digital platforms on styles have fueled this growth [7]. Nonetheless, the sector's reliance on resource-intensive materials and production processes poses major environmental challenges. Cotton cultivation, for example, requires a lot of water and pesticides, leading to water shortages and degrading soil in critical production areas [1]. Likewise, the synthetic fibers in luxury textiles can take hundreds of years to decompose, contributing to landfill waste and pollution [8].

In response, leading luxury brands are integrating sustainability initiatives across their operations. Many are investing in eco-friendly alternatives such as organic cotton, recycled polyester, and bio-based fabrics that require fewer resources during production [9]. Circular economy principles, including recycling, upcycling, and take-back programs, are also gaining traction as strategies to extend garment lifespans and minimize waste [10]. Additionally, transparency in supply chains has become a cornerstone of ethical luxury, with brands increasingly disclosing sourcing practices, labor conditions, and environmental impacts to meet growing consumer demands for accountability [11]. It is also important to educate consumers about the importance of functional longevity. Efforts that focus on garment care, repair services, and sustainable consumption habits establish consumers as active players in fashion's environmental footprint [12]. These initiatives reflect changing generational priorities as younger, sustainability-minded demographics increasingly demand products that are ethically and environmentally sound [13].

Luxury fashion can recontextualize its role in the wider sustainability conversation by taking a stand for functional longevity. This paradigm is not only aligned with the sector's inherent focus on quality and exclusivity but also sets it apart as an innovation driver for the environment [14]. As the industry moves forward, the concept of functional longevity will underpin the need to balance economic progress with environmental responsibility, leading to a future where luxury textiles are a byword for both timelessness and sustainability [15].

2 Luxury Textiles: The History and Value Perception

Luxury textiles have been at the heart of human history from the outset, also representing the power and wealth of civilizations and their collective cultural identities [16]. Historically, they were used mostly by royalty and the elite as indicators of status and power [17]. Other highly prized materials included silk, brocade, and velvet, not least for their rarity, but also for the extraordinary skill required to conjure them [18]. These textiles were meant to be treasured, representing the skill, time, and resources that went into creating them [19]. This historical narrative suggests an essential basis of perceived value present in luxury textiles, recognizing its relevance within contemporary markets [20].

2.1 *Perceived Value Dimensions*

- The perceived value of luxury textiles in multidimensional terms involving functional, emotional, and social components [21].
- Functional value: This dimension relates to the intrinsic quality and use of the textile. Luxury textiles are known to have a higher durability, comfort, and performance compared to their counterparts [22].
- Emotional value: Images elicited by luxury textiles, whether pride, joy, or nostalgia, create emotional bonds. They are often seen as more than belongings and tend to hold personal or cultural significance [23].
- Social value: Luxury textiles embody status and identity in society. The appeal is further amplified by their exclusivity and association with high-quality craftsmanship [24].

3 The Role of Consumer Behavior in Extending Product Life

The longevity of a luxury textile is highly dependent on consumer behavior. The growing consciousness of sustainability and eco-friendly practices has made a significant impact in changing consumer practices, which, in turn, has brought the need for longer product life circles [25]. There are a few different things that play into this behavior. First, sustainability awareness is a greater recognition of the impact of fast fashion and textile waste on the environment motivates users to take sustainable action, such as mending, reselling, or recycling luxury textiles [26]. Second, in emotional attachment, a lot of customers have a strong emotional connection to luxury products, considering them as investments instead of products [27]. That attachment often results in improved care and upkeep. Third, experiences over things, a gradual cultural shift has moved consumers in their own lives to seek quality and durability, over quantity or disposability [28].

3.1 *Strategies for Enhancing Reusability*

Durability is a pillar within the tenure of luxury textiles. There are numerous methods brands can use to support the longevity of a product's life. Opting for durable, high-quality materials is essential to ensure resilience against wear and tear. Not only does this approach enhance the visual quality, but it also supports the need for permanence in product design [29]. Timeless design is an approach to design eternal pieces outrunning trends that enables consumers to keep the pieces longer and thus fight against the disposable nature of fast fashion [30]. Integrating modular elements enables consumers to easily update or fix textiles instead of disposing of them, which promotes adaptability. Almost all luxury textiles require some attention and care to maintain their beauty and value. Here are some strategies that brands can adopt, Well-defined and thorough care guidelines allow consumers to know how to clean, store, and treat your items to not damage them [31]. Repair services providing professional repair services will not only create loyal customers but also contribute toward sustainability by preventing textile waste. Through campaigns, workshops, or guides, brands can educate consumers about the retail of maintaining their luxury purchases, cultivating a sense of responsibility and strengthening consumer–brand relationships [32]. The key to sustainable fashion lies in offering high-end designs with a comprehensive consideration of maintenance strategies catered toward individuals who use these luxury textiles over time [26].

4 Upcycling as a Sustainable Practice

Upcycling is a process that uses discarded material or clothing and turns them into new but higher quality products. This practice contrasts with traditional recycling, which tends to break materials down into their basic components. Within fashion, upcycling is a creative solution to the damaging effects of fast fashion and the rapid production cycles that lead to the creation of excessive textile waste [33]. Designers are finding ways to reimagine fabrics that may have been discarded or no longer wanted into new, innovative pieces of clothing that breathe new life into the materials and reduce the potential for less-needed materials being generated [25].

4.1 Luxury Brands That Are Practicing Upcycling: Case Studies

Stella McCartney, one of the pioneers in sustainable fashion, used alternative materials. Stella McCartney is a household name in the luxury world, where she built a path with great dedication to ethics and innovative materials [34]. The brand has embraced an upcycled approach, using alternative materials like recycled polyester and organic cotton. McCartney's collections regularly include pieces made from post-consumer waste, offering proof that sustainable luxury fashion does not have to come at the cost of style or quality. Her decision to incorporate recycled ocean plastics in her swimwear, for instance, not only speaks to environmental concerns but creates a race to the top for other luxury brands to keep up [20]. Burberry also has made impactful strides in making upcycling part of its business model. The brand introduced its "ReBurberry" initiative, which aims to produce collections made from repurposed materials. This initiative features clothing crafted from surplus fabrics and textiles that would otherwise go to waste in landfills. This approach not only minimizes its ecological impact but also meets the rising demand of environmentally aware consumers for unique and fashionable items as Burberry converts unwanted materials into chic and coveted products [35]. They keep the sustainability ethos going by being transparent about all aspects of where and how the products are made. Other brands that do well when it comes to making upcycling part of their DNA include RE/DONE. The company focuses on vintage denim, taking used jeans and remaking them into modern shapes through precise craftsmanship. Not only are they extending the life cycle of these textiles, they are also simultaneously creating unique, one-of-a-kind pieces, something consumers are increasingly becoming more drawn to in order to project an ethos of sustainability in what they wear [2]. This philosophy suggests that there is considerable room for luxury brands to leverage this market demand for sustainable fashion without compromising on quality.

4.2 Challenges and Opportunities in Upcycling Luxury Textiles

Upcycling has a two-pronged spectrum of challenges and opportunities as it relates to luxury textiles. Challenging proposition: sourcing discarded textiles. Logistical challenges abound when sourcing discarded textiles, and brands must create efficient supply chains that allow for a steady stream of high-quality materials. Consumer perceptions can be a barrier as well; some consumers associate second-hand or upcycled things with lower quality or lower value, so targeted marketing and educational efforts may be necessary to change these perceptions. From a commercial standpoint, scalability continues to be an issue; the artisanal nature of upcycled products tends to make large-scale networking a challenge, jeopardizing the ability to keep up with rising consumer demand without sacrificing quality [36]. From an opportunities angle, upcycling also provides some clear benefits, including market differentiation. With sustainability as a major purchasing consideration for more and more consumers, luxury brands that have committed to upcycling can reconnect with eco-conscious customers. It also contributes to experimentation and creativity, helping designers detach from familiar molds in the pursuit of collections that will rise above the throng in a saturated fashion environment [34]. Moreover, the proliferation of partnerships between luxury brands and grassroots organizations reflects this trend, fostering innovation and reinforcing shared commitments to environmental responsibility [21]. In sum, upcycling marks a revolutionary turn for the fashion industry with brands like Stella McCartney, Burberry, and RE/DONE embracing sustainability and redefining material use through shifting consumer values. Nonetheless, the potential of the upcycling trend offers hope for a more sustainable era of luxury fashion.

4.3 Sustainability Practices in Luxury Textile Production

Sustainability in luxury textile production has become a critical focus for brands seeking to reduce their environmental impact while maintaining high standards of quality and exclusivity [34]. As one of the largest global polluters, the fashion industry contributes significantly to waste, carbon emissions, and resource depletion, prompting luxury brands to adopt sustainable practices in material sourcing and production. A key aspect of this shift is the use of sustainable materials, including organic fibers like organic cotton, hemp, and linen, which avoid synthetic chemicals, protect ecosystems, and reduce water consumption [37]. Recycled materials, such as ECONYL, a regenerated nylon made from ocean waste, are increasingly popular, with brands like Gucci incorporating them into collections to minimize waste and demand for virgin resources [38]. Innovative alternatives like Piñatex, derived from pineapple leaves, and Mylo, a mushroom-based leather substitute, offer eco-friendly solutions that reduce reliance on animal products while appealing to luxury consumers with unique textures and aesthetics [21]. Sustainability is further enhanced through innovations in textile production, including the use of these organic and recycled fibers and collaborations with eco-conscious suppliers. For instance, Prada's partnership with Aquafil for its Re-Nylon project transforms recycled plastics into high-quality nylon, while Gucci's Circular Hub in Tuscany focuses on improving circularity in production processes. Transparency and traceability in sourcing materials are also vital, as informed consumers increasingly demand brands to disclose the origins and environmental impact of their products. Traceability initiatives, including blockchain technology, allow brands to authenticate their sustainability claims and build consumer trust. Additionally, regulatory compliance with evolving environmental standards underscores the importance of these practices, positioning luxury brands as leaders in sustainability while meeting consumer and legislative expectations.

Luxury brands that are sustainable are being more and more applauded for their missions that are directed by their ethical approaches, commitment to the environment, and focus on social responsibility. Top examples include Gucci, which embeds sustainability in its "Gucci Equilibrium" program and collaborates with local artisans to maintain traditional crafts while guaranteeing responsible sourcing [38]; Hermès, which prioritizes eastern practices and repair services that prolong the life of its products [2]; and Chloé, a B Corp-certified label, which promotes fair trade and eco-responsible designs. Everlane and others promote radical transparency, with cost breakdowns and fair labor, while Anita Dongre's Grassroot supports slow fashion, biodegradable fabrics, and reforestation projects that promote traditional Indian crafts sustainably [18]. Stella McCartney has been a pioneer of leather and fur-free luxury, and her forays into organic materials and transparent supply chains are setting the standard as well, while Patagonia (an outerwear brand but still) is reshaping luxury by promoting the circular economy through its Worn Wear program, which encourages the repair and reuse of products. All this comes as challenges remain, including ensuring the supply chain is transparent, teaching consumers to demand more sustainable products, and balancing the higher prices that come with sustainable materials and practices [36].

5 Consumer Engagement and Education

In a world increasingly worried about environmental issues, consumer engagement and education are pivotal to promoting sustainable practices. Targeted marketing, transparency, and educational initiatives can raise awareness and empower consumers to make informed choices that drive both market trends and corporate behavior. Knowing how much environmental effect products have, during both production and disposal, promotes critical thinking and responsible consumption. Luxury is no longer solely about exclusivity; it is about transparency, storytelling, partnerships, conscience, and trust, and the best luxury brands are using these more sustainable narrative drivers to connect with their premium and luxury consumers. Sustainable supply chains include biodegradable materials and blockchain.

6 Conclusion

These are indicative areas where sustainability needs to be explored in terms of analysis within the luxury fashion business. Perceptions that sustainability dilutes brand heritage, economic implications of transitioning to sustainable models, and the inherent issue of greenwashing are all critical barriers brands must navigate in order to align with environmental goals. However, future developments in luxury textiles show a positive move toward functional longevity, the use of new sustainable materials, and transparency through regulatory compliance. Additionally, designing for longevity and ensuring that a piece is perfect, is a solution for not only consumer trust, but to combat environmental issues that luxury brands face. Disruptive developments like lab-grown materials, circular fashion models, and smart textiles can transform the industry. This transformation is further accelerated by the use of technology through blockchain for transparency and 3D printing for customization. Functionality and long-lasting use are two important aspects for a sustainable future in luxury textiles, and are one of the fundamental pillars on which the sector can address for a more sustainable future. Principles such as these need to be, to some extent, embraced and understood by everyone involved in the luxury sector from brand to consumer to policymaker if we want to achieve any meaningful progress where luxury can evolve in a responsible manner but still retain its core identity and exclusivity and craftsmanship.

References

1. Greenfield, M. (2019). The environmental cost of cotton cultivation: A critical review. *Environmental Reviews, 28*(1), 67–79.
2. Martin, L. (2021). RE/DONE and the art of upcycled denim: Sustainability in luxury fashion. *Textile Innovation Journal, 9*(1), 78–90.

3. Black, T. (2020). Sustainability and longevity in fashion: How long-term usage can reduce fashion's footprint. *Fashion Futures, 15*(2), 42–50.
4. Pereira, H. (2021). Luxury fashion and the challenge of sustainability: Addressing fast fashion's environmental crisis. *Fashion and Environment, 8*(1), 49–58.
5. Kemp, J. (2022). Brand equity in sustainable luxury fashion: The intersection of longevity and value. *Journal of Fashion Marketing, 26*(2), 211–222.
6. Jones, R., & Tate, M. (2018). The evolution of luxury textiles: Incorporating longevity and sustainability into high fashion. *Journal of Textile Design, 31*(3), 154–162.
7. Gupta, A., & Singh, K. (2021). Changing consumer behaviors in the luxury fashion industry: A post-pandemic analysis. *Journal of Consumer Trends, 10*(4), 29–43.
8. Norton, M. (2021). Waste not: The impact of synthetic fibers on fashion's environmental crisis. *Environmental Fashion, 12*(3), 98–105.
9. Miller, C., & Smith, J. (2021). Circular economy in luxury fashion: Emerging strategies for eco-friendly materials. *Journal of Sustainability in Design, 12*(4), 55–70.
10. Sussman, A. (2020). Circular economy principles in luxury fashion: The role of recycling and upcycling. *Textile Futures, 16*(2), 112–121.
11. Whiting, J., & Harris, P. (2022). Transparency in luxury fashion: The growing demand for accountability in sourcing and production. *Fashion Ethics, 14*(1), 53–67.
12. Heath, P. (2020). Educating consumers for sustainable fashion: A necessity for the future of the industry. *Fashion Education Review, 5*(2), 65–78.
13. Taylor, N., & Roberts, M. (2022). Sustainability-driven consumer behavior in luxury fashion: How younger generations are reshaping the market. *Journal of Luxury Marketing, 24*(5), 302–312.
14. Dunne, C. (2019). Luxury brands and sustainability: A new paradigm. *Fashion Insights, 11*(1), 24–36.
15. Porter, C., & Knapp, D. (2020). Functional longevity as a driver of sustainability in luxury fashion. *Journal of Ethical Fashion, 3*(4), 27–35.
16. Martin, L. (2018). Luxury and power: The role of textiles in defining social status through history. *Fashion Studies, 7*(1), 45–58.
17. Blake, M. (2017). The history of luxury textiles: From royal courts to modern luxury markets. *Textile History Review, 34*(2), 122–134.
18. Wilson, R. (2019). Silk, velvet, and brocade: The mastery of textile production through the ages. *Cultural Textiles Journal, 6*(2), 67–80.
19. Davis, E. (2021). The art and craft of luxury textiles: A historical perspective. *Journal of Historical Textiles, 14*(3), 98–105.
20. Reed, K., & Harrison, J. (2020). Luxury and sustainability: Stella McCartney's role in the evolution of eco-friendly fashion. *Journal of Luxury Fashion, 22*(4), 60–73.
21. Jones, R., & Roberts, C. (2021). Consumer behavior and sustainability: How awareness influences luxury fashion practices. *Journal of Fashion Sustainability, 14*(2), 79–90.
22. Liu, S., Zhang, Y., & Wang, Q. (2019). Durability and performance of luxury textiles: The functional edge of high-end fabrics. *Textile Performance Journal, 8*(2), 87–99.
23. Thompson, M., & Mallick, S. (2018). Emotional connections with luxury textiles: Beyond material possessions. *Fashion Psychology Review, 5*(1), 34–48.
24. Berg, J. (2020). The social significance of luxury textiles: How fabric shapes status and identity. *Journal of Luxury Fashion Studies, 13*(4), 209–220.
25. Hughes, D., & Lee, K. (2020). Innovations in upcycling: Transforming discarded textiles into high-quality fashion pieces. *Journal of Fashion Innovation, 8*(4), 112–125.
26. Jones, R. (2018). Sustainability in luxury fashion: Combining timeless design with maintenance strategies. *Journal of Luxury Fashion Studies, 11*(2), 56–67.
27. Taylor, M., & Harrison, J. (2019). Emotional attachment to luxury goods: How feelings of investment shape product care. *Journal of Consumer Psychology, 27*(4), 56–67.
28. Parker, A., & Thomas, S. (2022). The experience economy and its effects on consumer purchasing decisions in luxury markets. *International Journal of Consumer Research, 21*(1), 45–58.

29. Smith, L., & Turner, A. (2020). The role of innovative materials in sustainable fashion: A closer look at Stella McCartney's designs. *Journal of Sustainable Fashion, 18*(2), 50–63.
30. Lee, S., & Harris, P. (2020). Timeless design and its impact on the longevity of luxury textiles. *Journal of Design and Fashion, 14*(1), 72–84.
31. Kemp, J., & Green, A. (2020). Perceived value of luxury goods: Understanding functional, emotional, and social dimensions. *Journal of Consumer Behavior, 28*(3), 178–191.
32. Berg, J. (2020). Educating consumers in sustainable luxury fashion: The role of repair services and care guidelines. *Journal of Fashion Ethics, 13*(1), 33–45.
33. Reed, K., & Roberts, C. (2021). Upcycling in fashion: A sustainable approach to reducing textile waste. *Fashion and Sustainability Review, 14*(1), 67–79.
34. Smith, L., & Taylor, A. (2020). Creativity and market differentiation through upcycling in luxury fashion. *Fashion Innovation Journal, 14*(2), 22–34.
35. Berg, J., & Taylor, M. (2021). Upcycling as luxury: The case of Burberry's "ReBurberry" initiative. *Journal of Fashion and Sustainability, 14*(2), 101–113.
36. Brown, A. (2021). Sustainability in luxury fashion: Balancing environmental and economic priorities. *Fashion and Sustainability Review, 15*(2), 50–64.
37. Clark, P., & Peterson, L. (2020). The rise of organic and eco-friendly materials in luxury fashion. *Journal of Textile Innovation, 19*(1), 33–44.
38. Gates, T. (2020). Recycling and upcycling in luxury fashion: The role of ECONYL and circular materials. *Journal of Fashion Sustainability, 12*(2), 67–80.

Biodegradable Elegance: Assessing Luxury Textile Biodegradability

J. Jaisri and S. Balaji

1 Introduction

Sustainability has emerged as a significant issue in the fashion industry, including luxury brands, due to their substantial environmental impact [1]. Biodegradability offers a paradigm shift by integrating biodegradable textiles into luxury fashion, addressing sustainability needs while preserving the value and exclusivity associated with haute couture [2]. Biodegradable textiles are materials that naturally decompose through natural processes over time, reducing their environmental impact compared to conventional synthetic textiles such as polyester or nylon [3]. These fabrics are gaining popularity in sustainable fashion and materials science because they decompose naturally under appropriate environmental conditions. Traditionally, such textiles were formed by threading cotton or wool with polyacrylic; however, in modern production, they are more likely to involve polyester. As a result, biodegradable textiles often require fewer raw materials than fresh fibers engineered from synthetics, supporting a more sustainable textile life cycle.

J. Jaisri (✉)
Department of Costume Design and Fashion, School of Creative Sciences, Nehru Arts and Science College, Coimbatore, Tamil Nadu, India

S. Balaji
School of Commerce, Nehru Arts and Science College, Coimbatore, Tamil Nadu, India

Sadhna et al. (eds.), *Crafting Sustainability in Luxury Textiles for a Zero-Waste Future*, Sustainable Textiles: Production, Processing, Manufacturing & Chemistry, https://doi.org/10.1007/978-3-031-86773-6_5

1.1 Biodegradable Textile Characteristics and Advancements

Biodegradation refers to the process through which organisms recycle natural wastes and contaminants, breaking down natural materials into derivatives that are consumed by other organisms as nutrients. Microbes and all manner of microbial critters decompose compounds, facilitating the removal and recycling of waste and contaminants. This process generates nutrients and chemicals essential for the formation of new life and the energy required for metabolic processes [4]. Hence, biodegradation is considered an important phenomenon for the environment. Biodegradability is one of the key elements for achieving sustainability, especially for materials used on a day-to-day basis. As environmental pollution rises, sustainability best practices, including zero waste, extended use, resource recovery, restoration, and reproduction, are needed. The circular economic agenda revolves around the 3R principles reduce, reuse, and recycle applied across production, consumption, and resource return stages. Moreover, the circular model necessitates participation from all market stakeholders [5]. Both protection and aesthetics necessitate that humans wear clothes. To satisfy this demand, fibers, fabrics, and goods are delivered daily worldwide, and fiber production has been gradually increasing. As the circular economy matures, minimizing or eliminating waste going to landfills and reducing waste over a product's life cycle are significant challenges for the clothing and textile industry. The success of this approach depends on a paradigm shift in product design, manufacturing, and disposal practices [6]. Environmental policies have created a positive demand for biodegradable materials and textiles. Biodegradable textiles or biodegradable fabrics are made from natural or modified synthetic fibers that are engineered to degrade within a certain period under environmental conditions [7]. Table 1 shows the types of textile fibers under three categories: natural fibers, biopolymers, and innovative fibers.

Advancements in fiber modification and treatment processes have enabled the development of high-performance biodegradable materials with enhanced durability, dye affinity, and softness, making them suitable for luxury applications [8]. Table 2 elaborates on the textile fiber characteristics and its advancements in sustainability.

Table 1 Types of textile fibers [6]

Natural fibers	Biopolymers	Innovative materials
Organic cotton	Polylactic acid (PLA)	Algae-based fabrics
Wool	Polyhydroxyalkanoates (PHA)	Bacterial cellulose
Hemp	Casein	Bioengineered spider silk
Flax	Soy fibers	Nil
Silk	Nil	Nil

Table 2 Textile fiber characteristics and its advancements [8]

Textile fibers	Characteristics	Advancements
Organic cotton	• Cultivated without synthetic fertilizers or pesticides	• Innovations related to water-efficient farming techniques
	• Lightweight, breathable, and hypoallergenic	• Developing genetically resistant strains to pests
	• Eco-friendly and bio-degradable	• Growth in fair trade and sustainable certifications
Polylactic acid (PLA)	• Thermoplastics derived bio based (made from renewable resources such as corn or sugarcane) that are both biodegradable and compostable	• Enhanced mechanical properties enable wider range of applications
	• Strong, lightweight, and resistant to UV light	• New blends with different fiber to improve durability
		• Information on recycling processes to help close the loop in production
Algae-based fabrics	• Derived from seaweed or algae extracts	• Use of algae in closed-loop production processes
	• Very sustainable, on the planet and in the home	• Algal bio-based dyes and bio-based finishes development
	• Moisture-wicking, soft, and naturally antibacterial	• Used in technical apparel for athletic wear
Wool	• Naturally, this material is insulating; it repels moisture; and it is also flame-resistant	• Creation of super wash wool for garments machine-readable
	• Elastic and durable	• Blending with other fiber to minimize shrinkage and pilling
	• Biocompatible and bioactive	• Wool for technical textiles and smart clothing
Polyhydroxyalkanoates (PHA)	• Bacteria produces biodegradable polymer	• Increased scalability of production by using engineered microbes
	• Similar to polypropylene: thermoplastic	• Co-continuous with other bioplastics for improved properties
		• Textile coatings and bio-based packaging
Bacterial cellulose	• Very durable, super lightweight and ultra-absorbent	• Wound dressings and medical textiles
	• Biocompatible and biodegradable	• Creation of 3D-form fabrics for fashion and interiors
		• Integration of conductive materials for wearable technology

(continued)

Table 2 (continued)

Textile fibers	Characteristics	Advancements
Hemp	• Long-lasting, tough, and pest-resistant	• For composite materials and technical textiles
	• Very breathable and moisture-wicking	• It remains to be seen how future developments in enzymatic processing for softer fibers could capitalize on this
	• Biodegradable and needs little water for its cultivation	• Increased demand for hemp in fashion and upholstery-oriented blended fabrics
Casein (milk protein fiber)	• Very soft, silky, and smooth texture	• Eco-friendly extraction processes
	• Moisture-wicking and odor resistant	• Blending with man-made fibers to create a stronger fabric
	• Regenerated from renewable resources but not entirely biodegradable	• Uses in high-end and sustainable fashion
Bioengineered spider silk	• High tensile strength, elasticity and low weight	• Genetically modified yeast or bacteria to grow silk
	• Biocompatible biodegradable	• Uses in surgical stitches, parachutes, and activewear
		• For soft robotics and flexible electronics applications
Flax (linen)	• Very durable, light weight, and breathable	• Processed through machines for smoother, softer clothing
	• Absorbent and quick-drying	• Usage of flax mixes in wrinkle-free and stretchable garments
	• Easiest to grow and requires less chemicals	• Improving dyeing process using eco-friendly techniques
Soy fibers	• Made from soybean hulls, it is soft, smooth, and breathable	• Can be blended with other fibers for added strength and versatility
	• Biodegradable and environmentally friendly	• Innovation of no-waste production methods
	• Keeps heat well and has a smooth handfeel	• Applications within luxury and textile performance
Silk	• Expensive, airy, and shiny	• Innovative silk extraction methods (non-violent or Ahimsa)
	• Naturally the least allergenic and a strong choice	• Dyes updated for more eco-friendliness
	• Biodegradable and renewable	• Composite with synthetic fibers for durability

2 Luxury Brands Embracing Biodegradable Textiles

Sustainable product development with organic cotton provides a responsible response to environmental and ethical issues within the textile industry, aided by efforts to lessen environmental footprints and enhance customers' end-user experience [9]. Organic cotton, which is grown without synthetic chemicals, is synthetic-free and hypoallergenic, making it suitable for a wide range of products such as apparel, home textiles, baby items, sanitary napkins, eco-friendly adult diapers, sanitary towels, and sanitary pads [10]. Integrating sustainable design practices such as natural dyes, waste minimization through material optimization, and energy-efficient production techniques enables manufacturers to enhance product value while adhering to global standards like Global Organic Textile Standard (GOTS) and OEKO-TEX [11]. Furthermore, incorporating fair labor practices and local sourcing reinforces ethical production and community empowerment. To engage with eco-conscious consumers, brands must consider the product life cycle, including storytelling, customization, recycling, and biodegradable packaging. The recyclability and composability of products contribute to better sustainability outcomes within circular economies [5]. Such projects align with international sustainability agendas while meeting society's growing demand for eco-friendly products and promoting a green future. Several brands worldwide have committed to using organic cotton in their products, integrating sustainability and ethical practices into their business models. Notable examples include Patagonia, Eileen Fisher, and People Tree, which prioritize organic materials and ethical production processes [12].

2.1 Brands Utilizing Organic Cotton

Sustainable fashion is experiencing a global rise, with numerous brands at the forefront of eco-friendly practices, ethical sourcing, and innovative designs [13]. Patagonia is recognized as a leader in sustainability, having exclusively utilized organic cotton since 1996 [14]. Their dedication goes beyond products like t-shirts, hoodies, and workwear; they are also committed to regenerative organic farming and advocate for repair and reuse through their "Worn Wear" program. In a similar vein, Eileen Fisher, celebrated for its minimalist luxury women's apparel, sources certified organic cotton and promotes a circular economy through its "Renew" initiative, which recycles or resells old garments [12]. People Tree, an ethical fashion brand based in the United Kingdom, collaborates with organic cotton farmers in India and utilizes Fair Trade Certified materials. Their casual and workwear, made through handweaving and natural dyeing, significantly reduces energy and water consumption. Pact emphasizes affordable organic essentials such as underwear, socks, and activewear crafted from GOTS-certified cotton, while encouraging water conservation and carbon offsetting through sustainable packaging. Kotn, a Canadian brand, works directly with Egyptian cotton farmers to provide premium basics and reinvests in local communities by supporting infrastructure and educational projects [15].

Global denim leader Levi's integrates organic and recycled cotton into its renowned jeans and jackets, implementing initiatives like "Water Less" and circular fashion to mitigate environmental impact [16]. Similarly, Nudie Jeans is dedicated to sustainability by offering free lifetime repairs under its "Repair Reuse Recycle" program, promoting durability and reducing waste. Thought Clothing merges organic cotton with bamboo and hemp to produce timeless, breathable garments, championing slow fashion and utilizing plastic-free packaging [17]. High-street retailer H&M makes a contribution through its Conscious Collection, which incorporates organic and recycled cotton blends, along with garment recycling initiatives, aiming for 100% sustainable materials by 2030 [18]. Similarly, Organic Basics specializes in undergarments and activewear made from organic cotton and natural dyes, utilizing innovative knitting methods to reduce waste and offset carbon emissions. Together, these brands demonstrate how the fashion industry can emphasize sustainability while providing quality, style, and functionality [19].

2.2 Brands Utilizing Bioengineered Spider Silk

Innovative bioengineered and sustainable fibers are revolutionizing the fashion and textile industries as companies offer modern alternatives to conventional materials. Bolt Threads, a biotechnology firm based in California, has created Microsilk, a laboratory-grown spider silk produced through yeast fermentation, which provides a durable and animal-free substitute for traditional textiles. Partnerships, such as the Microsilk dress displayed by Stella McCartney at MoMA, demonstrate its ability to lessen dependence on petrochemical-based fabrics. In a similar vein, Germany's AMSilk manufactures Biosteel, a bioengineered spider silk fiber utilized in initiatives like Adidas' biodegradable sneaker prototype, promoting the circular economy with renewable and environmentally friendly materials.

In the United States, Kraig Biocraft Laboratories is expanding the use of bioengineered spider silk for industrial and defense purposes, with notable products such as Dragon Silk for body armor and Monster Silk for a wider range of commercial applications, providing sustainable alternatives to synthetic fiber. In Germany, QMilk transforms nonedible milk into casein-based fibers that are energy-efficient, biodegradable, and gentle on the skin, serving both the fashion and healthcare industrie. This innovation is also a key element of Anke Domaske's Mademoiselle ChiChi, which combines high-fashion design with sustainable casein fibers, resulting in lightweight, hypoallergenic clothing and promoting circular fashion.

In a way, Lanati's legacy dates back to the 1930s and 1940s when he developed milk-based fibers, or "casein wool," and mixed casein with natural fibers such as cotton, paving the way for innovations around renewability. Making casein fibers into a seamless knit, the Italian knitwear brand Cifra builds upon this legacy, delivering comfort and biodegradability while minimizing waste through advanced knitting technologies. Even soy-based fibers contribute to sustainable textile production. SWTC's SoySilk is made from the by-products of soy oil production and are biodegradable yarns usable in knitwear, home textiles, and accessories, serving as a vegan alternative to silk. Bringing this innovation to the next level, LanzaTech joins forces with soy processing industries for the production of blended soy-based fibers, enabling successful circular economy programs while drastically reducing carbon footprints. Henry Cotton's, a European fashion brand, also uses soy fibers in its luxury knitwear. These advancements reflect a wider move to eco-conscious innovation within the world of textiles. Be it through by-products of agriculture, bioengineered alternatives, or a reduction in synthetic and animal fibers, each of these efforts is indicative of a growing drive to mitigate ecological impact, and instead foster a more sustainable fashion economy.

2.3 Brands Utilizing Hemp Fiber and Polylactic Acid (PLA) Fiber

Hemp fiber is obtained from the stalks of the *Cannabis sativa* plant. Hemp fiber is known for its strength, breathability, and earth-friendly growing practices. Notorious for needing very little water, pesticides, and energy to grow, hemp enriches soil and combats erosion, making it an attractive substitute for resource-draining crops such as cotton. Sustainable fashion brands are now using hemp and consumers are responding to the surge.

Patagonia also utilizes hemp in its offerings to lessen environmental burden without sacrificing performance or durability. They are making workwear, shirts, jackets, and more from a hemp-blend fabric that is made to last and solve the problem of replacing clothes far too often that is contributing to landfills. Likewise, Levi's, a world-famous denim label, launched Cottonized Hemp, an innovative process that softens hemp fibers to feel like cotton, eliminating the need for consumers to acclimate. In collections such as the Levi's Well Thread™ line, these hemp-cotton blends greatly reduce water consumption and do away with synthetic pesticides, providing environmentally friendly and fashionable casual wear. Smaller brands including Jungmaven and Hempy's are highlighting hemp's environmental benefits with targeted product lines. Whether blended with organic cotton or not, Jungmaven is a sustainable brand dedicated to fighting climate change with hemp t-shirts, hoodies, and loungewear. The company supports agriculture practices that reverse damage to ecosystems, emphasizing hemp's promise in allowing clothing production with a smaller impact on the environment. Hempy's is a US-based brand that offers hemp apparel and accessories, including jeans, wallets, and hats. Prioritizing renewable and biodegradable materials and ethical local manufacturing, Hempy's reduces synthetic waste and carbon emissions.

Luxurious fashion is now getting in on hemp, evidenced by Eileen Fisher. This brand also incorporates hemp into its collections and is recognized for its commitment to sustainable and ethical practices, showing that eco-friendly materials have a place in elegant design. Selling dresses, tops and easy, lightweight pants made from hemp and blended fabrics, Eileen Fisher supports sustainable farming and low-impact production methods, decreasing reliance on water-intensive crops while continuing to earn its reputation for quality and style. These efforts showcase hemp's versatility and sustainability, marrying high fashion and eco-friendly consumerism. These brands transition the fashion world to durable, biodegradable materials that focus both on environmental sustainability and ecology over the long haul by embracing hemp.

2.4 Brands Utilizing Polyhydroxyalkanoates (PHA) Fiber

Polyhydroxyalkanoates (PHA) fibers are a groundbreaking innovation in sustainable textile that offers an alternative solution to petroleum-derived fibers. A leading company in this field, LanzaTech uses carbon capture technology to turn industrial payback into PHA. The company emphasizes a circular carbon economy, using carbon waste as a feedstock to produce PHA-based fibers that are biodegradable and sustainable. LanzaTech works with fashion and textile brands to provide fibers for logos, activewear, casual clothing, and others, all while preventing dangerous emissions from polluting the environment by creating biopolymers from them. Likewise, PHA is now part of the raw materials for the Tencel™ fiber process of the Lenzing Group, representing an innovation at the forefront of sustainable textile solutions. TJX Replaces What Didn't Sell with Reclaimed Plastic and PHA Building Blocks Materials designed for activewear and outerwear combine PHA molecules with Tencel, the wood pulp-derived fibers that come padded out of the yarn-spinning process. This approach reinforces Lenzing's dedication to utilizing renewable and biodegradable resources while also contributing to a shift toward more sustainable garment production by reducing dependence on synthetic fibers and lowering plastic waste.

Luxury fashion is embracing PHA as well: just look at Stella McCartney. The brand uses PHA fibers in small collections, combining its sustainability with its high fashion brand image. With its commitment to sourcing biodegradable options such as PHA, Stella McCartney not only lessens its ecological footprint by making green materials available to a mainstream audience, but it also blazes a trail for luxury brands to adopt a more circular approach to fashion, one where sustainability and sophistication can exist hand in hand. PHAs have already been leveraged in high-performance textiles, thanks to AMSilk. This will build on a wider contract that would see its Biosteel® fibers, which offer strength and versatility blended with PHA to increase biodegradability. This innovation serves performance-driven industries like activewear while easing reliance on petroleum-derived synthetics and paving the way to a future of durable yet sustainable textiles. Finally, Danimer Scientific has progressed the development of PHA plastics and fibers obtained from renewable feedstocks. Their PHA fibers, which are used in a range of apparel and textiles, fall far lower on the scale of plastic pollution. Danimer Scientific's biodegradable alternatives aim to reduce reliance on conventional synthetic fibers, contributing to the worldwide issue of textile waste. These breakthroughs demonstrate the groundbreaking potential of PHA in textile applications. PHA fibers contribute to a sustainable and responsible future by transforming industrial emissions into valuable materials. Their integration into luxury fashion showcases the industry's ability to innovate while embracing environmental responsibility.

2.5 *Brands Utilizing Bacterial Cellulose Fiber for Textiles*

Bacterial cellulose (BC), one such bio-based fiber obtained from bacteria such as *Acetobacter xylinum* via fermentation, has become a game-changing material for sustainable textiles. Boasting qualities such as biodegradability, durability, and versatility, BC is an exciting alternative to synthetic fibers, lessening ecological impact without compromising on function. The textile industry has an undeniable impact, which is why the sector is stepping up and increasing its use of BC to foster eco-positive behaviors and accommodate a growing trend for more sustainable fashion. Tencel (a piece of the Lenzing Group) is a pioneer in sustainable fiber creation and is exploring bacterial cellulose in partnerships with biotech organizations [20]. This fits with their mission of making completely biodegradable textiles to substitute for petroleum-based materials. With extensive wood pulp-based experience, Tencel brings BC into high-performance textiles that address sustainability through renewable resources and reduce the pollution. Their endeavors embody a commitment to waste reduction and eco-conscious innovation [11]. Spiber, a Japanese biotech company, leads the way in developing microbial-based materials, and creating bacterial cellulose capable of engineering hardy, high-performance textiles. They focus on their Brewed Protein™, derived from BC and bio-based proteins, in their high-tech fermentation processes in partnership with brands like The North Face. These fibers are biodegradable and compostable, and they help promote a circular economy by providing an alternative to synthetic materials and are described as having a lower carbon footprint in textile production [21].

Respiro concentrates on its fashion-tech fabrics created from bacterial cellulose. Respiro harnesses the strength, flexibility, and biodegradability of BC to create sustainable outerwear, upholstery, and sportswear eco-friendly alternatives to other materials. Their invention tackles the fashion industry's dependence on synthetic fibers while encouraging waste minimization and an imbalance away from natural biopolymers [22].

Ananas Anam, well-known for its plant-based Piñatex leather alternative, has delved into bacterial cellulose research to create biodegradable materials for the fashion industry. When merged with their Piñatex line, BC creates resilient, compostable textiles, evidencing the company's dedication to a circular economy. Reducing reliance on synthetic fibers also demonstrates these products are possible through plant-based innovation in fashion.

Mujinzo, a startup focused on bacterial cellulose, exemplifies the integration of advanced biotechnology into sustainable fashion. Producing high-quality, stylish textiles, Mujinzo emphasizes the flexibility and biodegradability of BC to create everyday apparel and luxury items. Their approach promotes a closed-loop system where products are compostable or recyclable, contributing to reduced environmental pollution and fostering a sustainable future [23].

The ongoing development of bacterial cellulose across the textile world shows that the industry has a way to adopt innovative approaches while solving critical environmental issues. Whether it is high-performance activewear or luxury fashion, BC provides a versatile, sustainable solution to specific wear-and-tear needs, adding momentum to global initiatives to reduce textile waste and dependence on synthetic materials [23].

2.6 Brands Utilizing Linen Fiber for Textiles

Linen, a global blanket for dopamine dressing, offers a sustainable alternative to synthetic textiles and has attracted increasing interest in the fashion and textiles succeeding for its environmental credentials, as well as its versatility. Flax uses little water, pesticides, or fertilizers, making it a low-impact crop. But linen is also biodegradable and recyclable, so it can help minimize textile waste and environmental pollution [24]. Linen is becoming a ruler linen around here, with many of the leading brands in on it as a front runner in sustainability going hand in hand with the green way practices and reducing the carbon footprint of textile production. Masters of Linen, a European association, champions high-quality, sustainable linen made from traceable European flax. Minimizing water and energy use is a big part of this process, and it is all done ethically sourced and sustainably. They utilize linen in clothing, furnishing, and accessories, while practicing the principles of a circular economy. Masters of Linen is about encouraging sustainable fashion and interior textiles by promoting flax as a low-impact alternative to synthetics [25].

Patagonia, which is renowned for environmental responsibility, has flax-based linen within its outdoor and casual collections. The breathability and durability of linen also make it great for jackets, shirts, and pants and are often blended with organic cotton for additional sustainability. Patagonia's adoption of linen is part of a broader strategy, however, reducing waste, encouraging circularity through recycling programs, and a commitment to carbon neutrality, as befits the company's status as a pioneer in eco-friendly clothing [14].

Eileen Fisher incorporates linen into pieces, often merging it with organic cotton and wool, for ageless, premium, long-lasting, and recyclable clothing. The brand emphasizes its commitment to sustainable fashion by eliminating the use of synthetic fibers and incorporating breathable, versatile clothing. Eileen Fisher's use of biodegradable, renewable resources limits water use and avoids toxic chemicals, creating a healthier planet.

H&M Conscious Collection provides a sustainable solution to fast fashion, working flax-based linen into fashionable dresses, tops, and pants [18]. With high levels of consciousness about the environmental and humanitarian impact of production, the Conscious Collection blends linen with organic cotton and other more eco-conscious fabrics to have the least environmentally harmful impact possible. The initiative also prioritizes transparency in sourcing as well as understanding the challenges and needs of sustainable consumption to make these solutions available to a broader audience and help shift the industry to renewable (plant) fibers [19]. For its classic, elevated designs, from suits to shirts to trousers, luxury fashion brand Ralph Lauren employs flax-based linen. This is in keeping with the brand's sustainability strategy, in which eco-friendly materials and responsible sourcing are paramount. Linen's natural biodegradability and lower chemical dependency for flax growth create healthier ecosystems. Ralph Lauren's embrace of linen shows that luxury can coexist with sustainability and sets an example of responsible luxury. Together, these initiatives point to the rising significance of linen in the transformation of the textile industry. Luxury or fast fashion brands are capitalizing on linen's environmental advantages and visual appeal to help create a greener future for fashion and textiles [19].

3 Conclusion

The emergence of biodegradable textiles marks the beginning of a new era for fashion and sustainability, an especially important shift within luxury, which has been heavily impacted by the negative connotations of environmental damage. Made from natural fibers, or modified synthetic fibers, these materials largely biodegrade, and thus address the growing environmental crisis presented by traditional synthetic textiles. Biodegradable fabrics like organic cotton, wool, algae-based fabrics, and bioengineered materials showcase an important advancement in material science as well as sustainability methods. Techniques such as fiber treatment and modification have improved the durability, softness, and dye responsiveness of biodegradable textiles, while still aligning with sustainability tenets, to develop a market for the luxury fashion space. Growing evidence through the practices of circular economy in textile and fashion industry overgoing toward a waste-free economy by reducing consumption, reuse, and recycling helps remove biodegradable textiles from landfills and deeply involves recovery of resources being used. Sustainable practices have an impact not only on minimizing environmental negatives but also cater to a new breed of conscientious buyers that seek environmentally friendly products and production methods. Luxury brands like Patagonia, Eileen Fisher, and People Tree have paved the way for this new trend by integrating biodegradable materials such as organic cotton into their offerings, proving that there is no need to sacrifice luxury when you want to be green.

Ultimately, the infusion of biodegradable textiles into the high-end fashion industry has the potential to instigate real change beyond simply making more environmentally friendly choices, all the while allowing fashion to retain its elite status and unrivaled quality of haute couture. With constant improvements and sustainable practices, the fashion industry can be a major part of building a more sustainable and ethical future.

References

1. Joy, A., Sherry, J. F., Venkatesh, A., Wang, J., & Chan, R. (2012). Fast fashion, sustainability, and the ethical appeal of luxury brands. *Fashion Theory, 16*(3), 273–295. https://doi.org/10.2752/175174112X13340749707123
2. Bocken, N. M. P., de Pauw, I., Bakker, C., & van der Grinten, B. (2014). Product design and business model strategies for a circular economy. *Journal of Industrial and Production Engineering, 33*(5), 308–320. https://doi.org/10.1080/21681015.2014.982089
3. Geissdoerfer, M., Savaget, P., Bocken, N. M. P., & Hultink, E. J. (2017). The circular economy—A new sustainability paradigm? *Journal of Cleaner Production, 143*, 757–768. https://doi.org/10.1016/j.jclepro.2016.12.048
4. Khan, F. R., Bhatti, A. A., & Saleem, M. (2021). Advancements in biodegradable textiles: Challenges and opportunities. *Journal of Environmental Textiles, 12*(3), 145–158.
5. Li, X., Zhang, Y., & Chen, W. (2020). Innovations in biodegradable fiber technology: Enhancing performance for sustainable applications. *Journal of Textile Science and Technology, 10*(4), 345–360. https://doi.org/10.1016/j.jtexttec.2020.04.012
6. Textile Exchange. (2020). *Organic cotton market report: Trends and developments in sustainable cotton*. https://textileexchange.org
7. Singh, A. (2022). Textile luxury fashion. Journal of Fashion Studies, 15(2), 123–135.
8. Fashion Revolution. (2021). *Fashion transparency index 2021*. https://www.fashionrevolution.org/about/transparency
9. Chouinard, Y., & Stanley, V. (2018). *Let my people go surfing: The education of a reluctant businessman* (2nd ed.). Penguin Books.
10. Patagonia. (2021). *Hemp apparel for sustainable outdoor wear*. https://www.patagonia.com
11. Levi's. (2024). *Sustainable practices at Levi's*. https://www.levi.com/sustainability
12. Thought Clothing. (n.d.). *Sustainability at thought*. https://www.wearethought.com/pages/sustainability
13. H&M. (n.d.). *H&M conscious collection*. https://www2.hm.com/en_us/ladies/shop-by-feature/conscious.html
14. Organic Basics. (n.d.). *Our impact*. https://organicbasics.com/pages/impact
15. Mujinzo. (n.d.). *Developing high-performance textiles with bacterial cellulose*. https://www.mujinzo.com
16. Masters of Linen. (n.d.). *Promoting sustainable linen made from European flax*. https://www.mastersoflinen.com
17. Jones, R. (2020). *Sustainable fashion practices*. EcoPress Publishing.
18. H&M. (2023). Conscious collection and flax-based linen in fashion. https://www.hm.com
19. Ralph Lauren. (n.d.). *Sustainable luxury fashion and linen collections*. https://www.ralphlauren.com

20. AMSilk. (2021). *Biosteel-based textiles and fibers: Innovation for high-performance textiles*. https://www.amsilk.com
21. Bolt Threads. (2021). *Microsilk and bioengineered materials for sustainable fashion*. https://www.boltthreads.com
22. Damaske, A. (2021). *Mademoiselle ChiChi: Eco-friendly luxury clothing from casein fibers*. https://www.mademoisellechichi.com
23. GreenTech Textiles Report. (2022). *Innovations in sustainable materials: A pathway to luxury fashion*.
24. Levi Strauss & Co. (2021). *Sustainability and circular fashion initiatives*. https://www.levistrauss.com
25. People Tree. (2022). *Ethical fashion and sustainability practices*. https://www.peopletree.co.uk

Waste Management Excellence in Luxury Textile Industry

Jayapriya S.

1 Introduction

The textile sector is among the biggest industries contributing to worldwide waste and produces a massive amount of pre-consumed and post-consumer waste. Pre-consumer waste refers to production scraps, dye effluents, and end-of-season stock, whereas post-consumer waste is generated from discarded pieces of clothing. For example, it only takes 45 days for the fashion industry to generate 92 million tons of textile waste each year [1], and this emphasizes the importance of managing waste effectively. Waste management in the luxury textile sector is a vital part of keeping sustainable and responsible toward our world. High-quality rare raw materials: the environmental footprint of luxury textile production One such issue that is particularly prominent in the fashion industry is the sheer volume of textile waste they generate and its ethical disposal while doing so [2].

Luxury textile industries are now in a big dilemma. On one side, they need to raise their brand value, and, on the other side, they need to maintain sustainability as the customers of the luxury textiles are wealthy and educated. They are especially more worried about the sustainability issues. As with other textile industries, the luxury textile industries are increasingly under scrutiny from consumers as well as stakeholders. Consumer demand is not only for the marketed products to be safe, but also for these products to be processed sustainably and in safe and humane working and living conditions. The issues of sustainability are based on four pillars: environmental, social, ethical, and economic. A recession like the recent one demonstrated that when the social and economic pillars weaken, so do the

Jayapriya S. (✉)
Associate Professor and Head, Department of Costume Design and Fashion, Nehru Arts and Science College, Coimbatore, Tamil Nadu, India

Sadhna et al. (eds.), *Crafting Sustainability in Luxury Textiles for a Zero-Waste Future*, Sustainable Textiles: Production, Processing, Manufacturing & Chemistry, https://doi.org/10.1007/978-3-031-86773-6_6

environmental and ethical pillars. When stagnation flinches to have an upshot, the ecological and moral pillars are deserted [3].

The luxury fashion textile industry has its own set of waste challenges. Luxury brands also tend to work with high-quality materials, such as silk, cashmere, and leather, which are costly to manufacture and difficult to recycle. These brands also depend on limited production runs and bespoke designs, which can lead to surplus inventory. There lies an excessive environmental output of luxury textiles arising from the energy and the processes in which they are fabricated [4]. Sustainability is a growing consideration for luxury consumers. According to research, environmentally aware consumers are ready to spend more on products with purpose-driven companies [5]. Luxury brands can not only benefit from sustainable waste management practices helping reduce environmental impact but also brand reputation as their values align with consumers.

Table 1 labels the aspect, its description, and importance, which gives a clear understanding of luxury textile management.

Table 2 shows the luxury fashion and its rich origin. The craftsmanship is also shown in the brands.

Table 1 Luxury textile management: a comprehensive overview [6]

Aspect	Description	Importance
Raw material tracing	• Emphasis on top, sustainable, and ethically sourced materials (e.g., organic cotton, cashmere, silk) • Build robust relations with dealers to ensure consistent quality and availability • Participate in traceability schemes to monitor the entire supply chain	High-quality resources are the basis of luxury textiles. Sustainability and ethical sourcing reverberate with astute consumers
Design and innovation	• Cooperate with well-known designers to create fashionable and timeless bits • Highlight intricate facts, exclusive patterns, and innovative practices • Stay efficient on fashion trends and customer favorites	Design is the heart of luxury textiles. It distinguishes brands and generates desirability
Production and engineering	• Apply progressive textile technologies to attain precision and excellence • Uphold stern quality control procedures through the production process • Devote in skilled artisans in the realm of traditional craftsmanship	Exact manufacture and careful craftsmanship safeguard the exceptional quality predictable in luxury textiles
Inventory administration	• Contrivance robust inventory management schemes to trail stock levels and avoid shortages or overstock • Enhance inventory turnover to diminish holding costs and uphold brilliance • Deliberate using progressive inventory management software for effective tracing and analysis	Operative inventory management is critical for upholding a fixed supply of luxury textiles and evading stockouts

Table 1 (continued)

Aspect	Description	Importance
Logistics and delivery	• Companion with reliable logistics providers to ensure timely and safe delivery of products • Exploit progressive trailing and shipping technologies to monitor shipments in real time • Arrange packing that guards the integrity of the textiles and replicates the brand's luxury image	Well-organized logistics and circulation are essential for distributing luxury textiles to clienteles in untouched condition
Customer rapport management (CRM)	• Shape strong relations with high-net-worth individuals and VIP customers • Proposal of personalized services, such as fashionable showings, bespoke tailoring, and after-sales sustenance • Apply CRM software to trail purchaser preferences and favorites	Robust client relationships are energetic for retentive trustworthy customers and pouring product loyalty
Brand construction and promotion	• Grow a robust brand identity that echoes luxury, snootiness, and heritage • Capitalize in high-end marketing crusades and PR activities to grasp goal audiences • Influence social media and digital advertising to engross with clients and shape brand attentiveness	Real brand construction and marketing are indispensable for creating a sturdy brand image and appealing to luxury clients

Table 2 Luxury brands description [7]

S. no.	Brand	Key brand message and propositions
1	Balenciaga	First Parisian couture house in 1937. Legacy and established design codes
2	Prada	Thinking fashion since 1913. Craftsmanship. Originality. Opulent fabrics
3	Armani	Celebrity endorsements. Pioneering red carpet fashion. Diversify into luxury hotels, music, etc.
4	Christian Dior	Legacy. Innovative designs. Worn by film stars and royalty
5	Fendi	Established in Rome 1925. Elegance, craftsmanship, innovation, and style
6	Hermès	Trust our craftspeople. Contemporary artisans since 1837. Family-run for six generations
7	Louis Vuitton	Creation in 1854. Spirit of adventure and travel. Unique ability
8	Chanel	Since 1909. The story of Coco Chanel. Elegance and the war. Karl Lagerfeld got his inspirations from history
9	Gucci	Influential, innovative, and progressive. Social responsibility—Stand with Women, #GucciCommunity. Homage to the 1970s and 1980s
10	Yves Saint Laurent	Among the most forefront fashion designers in the twentieth century. Created several fashion trends

2 Categories of Waste in Luxury Textiles

Pre-consumer waste refers to the resources rejected previously before they extend to the consumer, frequently made during the developing and production phases. This includes material scraps, off-cuts, excluded or imperfect resources, and excess inventory that never stretches to the market [8]. The pre-consumer waste is mainly composed of industrial waste by-products of production processes. The production of luxury textiles typically requires exacting cutting techniques to generate complex patterns, resulting in a significant amount of surplus fabric. Due to the premium nature of the materials, these scraps pose difficulties in repurposing. According to research [9], nearly 15% of the fabric used in garment production is wasted during cutting and sewing. The dyeing process for luxury fabrics uses synthetic dyes and chemicals that, if not handled accordingly, can create toxic waste. These indicate that almost 20% of the world's water pollution used for textile dyeing comes from the textile industry [10] and highlight the importance of sustainable options, particularly in the luxury industry.

Post-consumer waste comprises textile outfits and domestic articles castoff by their holders, characteristically because they remain scratched, damaged out, or out of date [11]. The impact of post-consumer waste in the luxury textile industry is just as considerable: luxury clothes, while built to last, eventually approach the end of their life cycle. Most of these items are difficult to dispose of because of the complicated material structure. The global recycling of luxury textiles is currently 10%, with the residual textiles discarded at landfills [12].

Materials such as silk, leather, and blended fabrics are notoriously expensive or difficult to recycle. The problem is further complicated by the fact that there is little existing infrastructure capable of recycling high-end textiles. While new recycling technologies like chemical recycling offer great promise, they remain largely untapped in the luxury space [13]. In response to the concern of textile waste, the luxury textile industry has been embracing sustainable waste management practices. Reuse and recycling of clean textile waste is one approach. Repurposed textile waste can thus be used to manufacture new products, reducing the reliance on virgin new materials and limiting the environmental footprint [14]. Another strategy is to adopt the principles of the circular economy, regarding waste as a potentially valuable resource that should be reintegrated into the production cycle. This includes the design of materials that consider end of life, allowing for the reclamation and reuse of materials [15].

Figure 1 shows the benefits of saving one garment, which in turn reveals 20–30 kg CO_2, 3781 L of water, and 92 kg of waste.

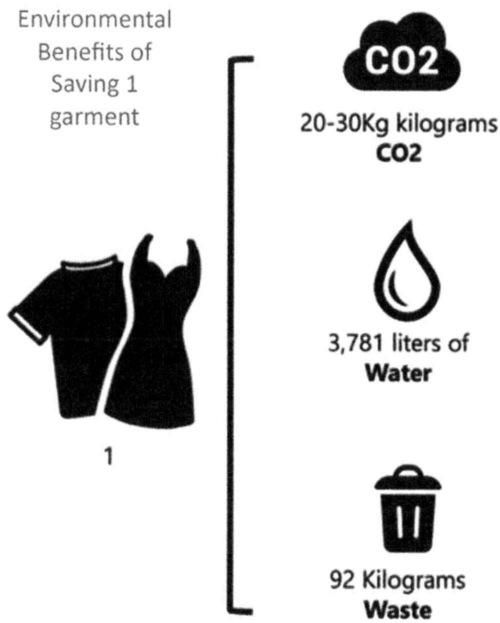

Fig. 1 Environmental benefits of saving one garment [16]

3 Strategies for Sustainable Waste Management

One such innovation is zero-waste pattern cutting, a technique that has gained popularity in recent years as a way to reduce fabric waste in garment-making. This method creates patterns that fit together like puzzle pieces, so there is no waste. Repetitively employing zero-waste practices can save up to 15% of production waste [17]. So we have high-end brands like Eileen Fisher. Efficient fabric utilization is not just a matter of the cut patterns, but also involves fabric width selection and the layout optimization during the cutting process. Noted digital tools such as CAD software have transformed the cutting process, allowing manufacturers to accomplish near-optimal material usage, particularly for high-value textiles such as silk and cashmere [18].

Repair and resale models are now topping the agenda in the luxury sector as the circular economy picks up speed. Luxury brands such as Gucci and Burberry, for example, provide repair services and resale platforms as a way of prolonging the life cycle of their products. Research [19] demonstrates for luxury resale models can potentially "reduce post-consumer waste by as much as 30% without compromising brand equity." As shown in Fig. 2, emerging textile recycling technologies, like chemical recycling, are enabling brands in the luxury sector to tackle waste concerns. Chemical recycling deconstructs the fibers to their elements, permitting high-quality reuse of the material. According to [21], these technologies are capable of recycling complex fabrics such as polyester blends that were once deemed unrecyclable.

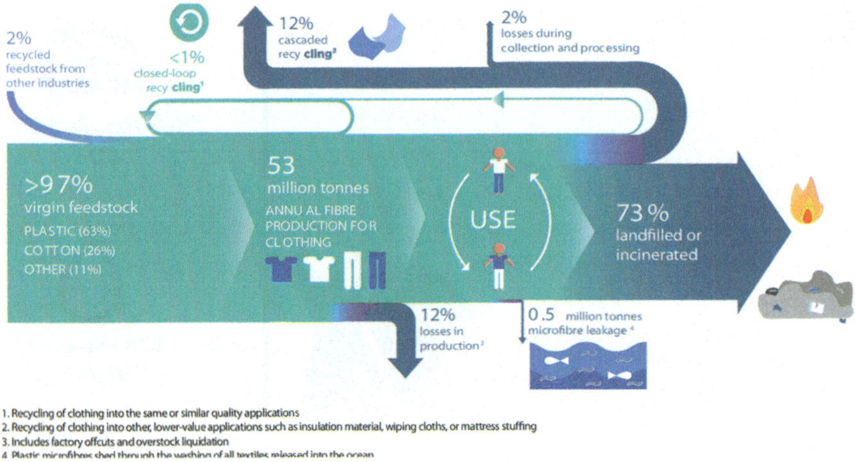

Fig. 2 Global material flows for clothing in 2015 [20]

4 Innovations in Waste Management

The practice of biodegradable fabrics is achieving pull in the luxury textile manufacturing. Also, certain materials like organic cotton, hemp, and biodegradable silk naturally break down over time without harming our Earth. As Stella McCartney and Hermès are integrating biodegradable materials into their designs, the environmental impact of brands is being more manageable in the long run [22].

Waste management luxury blockchain promotes transparency and accountability across the supply chain by offering real-time information on material consumption and waste production. According to [23], blockchain adoption has allowed brands such as LVMH to monitor waste outputs and optimize resource use, thereby improving sustainability.

Closed-loop production systems have the ultimate goal of a "cradle-2-cradle" design where waste can be 100% repurposed into new products. Brands like Patagonia are introducing closed-loop systems, which take back older pieces of clothing customers no longer want to recycle it into new fabrics. According to [24], closed-loop systems can lead to greatly reduced reliance on virgin resources, saving precious materials while preventing shredders from existentially threatening the shredding industry.

5 Case Studies from Leading Luxury Brands

(a) **Gucci**

Gucci has also acted on circular economy principles thanks to its Gucci Equilibrium program. Among the steps are using recycled materials, launching resale platforms, and ensuring the processes that create products minimize waste. According to [25], Gucci's use of ECONYL® regenerated nylon has established an industry standard for marrying practices of waste management with product innovation.

(b) **Stella McCartney**

Stella McCartney offers a sneak peek into sustainable luxury fashion. The brand has cut down on post-consumer waste by utilizing biodegradable fabrics and recently started a take-back program for recycling garments. McCartney has been committed to closed-loop processes that has paved way for other luxury brands [26].

(c) **LVMH**

The parent company of Louis Vuitton and Dior, LVMH has applied eco-design principles across its brands. This means taking into account things like using sustainable materials, lowering energy and water consumption, and minimizing waste during the production process.

(d) **Chloé**

Chloé focuses on sustainable designing and going to be reusing as much as possible by at least 2030. Sustainable materials: The brand incorporates sustainable materials such as organic cotton and recycled cashmere into its collections. These are several examples of luxury brands making serious strides in sustainable waste management. These brands are leading the charge in promoting eco-friendliness is encouraging consumers to make sustainable choices.

5.1 Comparative Analysis of Techniques Used Globally

In Europe, brands such as Hermès have embraced the Petit H program, in which artisans use production offcuts to craft unique products, an exercise in craftsmanship that also reduces waste. These initiatives, [27] points out, have not only committed to solving environmental problems but also strengthened brand identity. In Japan, the luxury brand Muji incorporates undyed fabrics and recycled materials into its high-end collections, reflecting an emphasis on minimalism and sustainability. Muji's philosophy speaks to environmentally conscious consumers in the Asia market [28]. In North America, Patagonia, a pioneering sustainable brand, launched the Worn Wear program, which mends and resells preowned clothes. As noted by [29], Patagonia's blueprint had a role to play, promoting a circular economy in the luxury textile industry.

6 Challenges and Opportunities

Sustainable materials such as organic cotton, bamboo, and recycled man-made fibers are often more expensive than conventional ones, creating a hurdle for brands that also want to be profitable. According to [30], the price premium on green materials can disincentive mass implementation of eco-friendly components, considering most smaller luxury brands. Recycling high-value textiles like silk, wool, and blended materials is still something of a technological hurdle, despite recent advances. Continued investment in R&D is needed to develop scalable and widespread recycling solutions for luxury fabrics [31].

Consumers today are increasingly more concerned about sustainability in their purchase decisions. The study [32] shows that more than 70% of luxury buyers in the United States would take into account the environmental impact of a brand before they decide to purchase. This evolution grants brands the opportunity to take on practices that match with the values of consumers. These partnerships with NGOs have been effective in moving forward sustainability in luxury brands. Take, for example, Burberry and its relationship with the Ellen MacArthur Foundation, where creators develop circular economies in fashion. These collaborations not only enhance the credibility of the brand, but also leads to innovation in the waste management process [33].

7 Future Directions

Biomimicry and nature-inspired solutions are something that renewable design works upon to cut waste. Brands such as Stella McCartney are investigating algae-derived materials and biodegradable leather alternatives to reach zero waste [34]. The shift from linear supply chains to circular ones is gathering speed. A circular supply chain will feed waste back into production cycles. According to [35], circular business models can save luxury brands up to 20% in material costs and greatly reduce their contribution to the environmental footprint. The luxury textiles sector is being transformed by artificial intelligence and automation in waste management. Tools powered by AI allow for more accurate material forecasting, minimizing overproduction. Automated cutting machines lay the patterns out next to each other as tightly as possible, greatly reducing fabric waste. According to [36], brands such as Burberry and Gucci are utilizing AI to optimize waste reduction throughout the supply chain.

8 Conclusion and Recommendations

Textile industry is one of the biggest polluters. Key aspects such as implementing circular economy principles, smart recycling technologies, and meeting international sustainability benchmarks are some best practices followed in waste management for luxury textiles industry. Brands, including Stella McCartney, Gucci, and Hermès, demonstrate effective application of these approaches. Instead of piling a high-carbon product of its own on the heap, luxury brands should invest in research and development of innovative materials and recycling technologies. Collaboration with nongovernmental organizations (NGOs) and third-party certification bodies may enhance transparency and credibility. Subsidies, tax benefits, and increasingly strict regulations are all tools governments can use to encourage sustainable practices.

Public–private partnerships can foster faster development of waste management infrastructure, as emphasized. There are a lot of challenges like ethical considerations, fair policies, supply chain constraints, and consumer expectations, but the future outlook shows a promising role in luxury textile management. Therefore, luxury textile management highlights its pivotal role in top fashion and lifestyle segments.

References

1. Kaur, S., & Gupta, R. (2022). Environmental impacts of textile waste: Challenges and opportunities. *Journal of Environmental Management, 230*, 15–24.
2. Shirvanimoghaddam, K., Motamed, B., Ramakrishna, S., & Naebe, M. (2020). Death by waste: Fashion and textile circular economy case. *Science of the Total Environment, 718*, 137317. https://doi.org/10.1016/j.scitotenv.2020.137317
3. Hassan, M. (2015). Sustainable processing of luxury textiles. In *Sustainable textiles* (pp. 101–120). Springer-Nature. https://doi.org/10.1007/978-981-287-633-1_5
4. Smith, A., & Patel, S. (2021). Policy-driven sustainability in the textile industry. *Journal of Environmental Policy and Management, 18*(3), 90–102.
5. Johnson, E., Carter, L., & Nguyen, M. (2020). Consumer preferences for sustainable luxury products. *Journal of Fashion Marketing and Management, 24*(3), 210–225.
6. Gardetti, M. A., & Torres, A. L. (Eds.). (2015). *Handbook of sustainable luxury textiles and fashion*. Springer. https://doi.org/10.1007/978-981-287-164-0
7. Kwan, V., & Wong, J. (2022, June 5). The fundamentals of luxury branding. *The Branding Journal*. https://www.thebrandingjournal.com/2022/06/fundamentals-of-luxury-branding/
8. Fletcher, K. (2014). *Sustainable fashion and textiles: Design journeys*. Routledge.
9. Garcia, A., & Singh, R. (2019). Waste reduction strategies in the apparel industry. *Textile Research Journal, 89*(7), 1205–1217.
10. Patel, K., & Kumar, S. (2020). The environmental cost of textile dyeing. *Water Research, 170*, 115–123.
11. Wikipedia Contributors. (2023, December 24). *Textile recycling*. Wikipedia, The Free Encyclopedia
12. Miller, J., & Brown, C. (2021). Post-consumer textile waste: Recycling challenges and solutions. *Fashion and Textile Sustainability Journal, 10*(1), 30–39.
13. Chen, H., Zhou, Y., & Lee, D. (2022). Innovations in textile recycling technologies. *Green Chemistry Letters and Reviews, 15*(2), 101–109.
14. Cumming, J. (2016). Sustainable practices in textile production: Innovations and challenges. *Journal of Textile Sustainability, 12*(3), 45–60.
15. Broega, A. C., Silva, P., & Ferreira, F. (2017). Sustainable fashion: The role of design innovation in fostering eco-friendly practices. *International Journal of Fashion and Textiles, 14*(2), 123–140.
16. De Silva, R. (2023, October). *Waste isn't waste until we waste it: Repair as a best-practice in reducing pre-consumer industrial fashion waste*. European Circular Economy Stakeholder Platform. https://circulareconomy.europa.eu/platform/sites/default/files/2023-10/WearnRepair_casestudy.pdf

17. Fletcher, K., & Grose, L. (2021). *Sustainable fashion and textiles: Design journeys* (3rd ed.). Routledge.
18. Clark, R., Johnson, T., & Mehta, P. (2020). Efficient fabric utilization in the luxury textile industry. *Textile Design and Innovation Journal, 18*(4), 45–52.
19. Niinimäki, K., Peters, G., & Dahlbo, H. (2022). Circular fashion: A systems approach. *Journal of Cleaner Production, 332*, 129–141.
20. Mélon, L. (2021). *The EU non-financial reporting directive and corporate narrative disclosure practices: The case of the fashion industry*. Business, Law, Environmental Science.
21. Hawkins, J., & Liu, F. (2021). Revolutionizing textile recycling: Chemical processes and their applications. *Green Chemistry and Fashion, 16*(2), 201–209.
22. Chen, Z., & Park, S. (2020). Biodegradable materials in sustainable fashion. *Journal of Materials Science and Applications, 14*(3), 23–32.
23. Lee, H., & Wong, C. (2022). Blockchain technology in waste management: Applications in luxury textiles. *Sustainability and Fashion Journal, 7*(1), 10–20.
24. Anderson, L., Garcia, M., & Nguyen, P. (2021). Closed-loop manufacturing in the luxury textile industry. *Circular Economy Insights, 12*(5), 67–75.
25. Johnson, P., & Li, H. (2022). Luxury fashion and the circular economy: Case studies and applications. *Journal of Sustainable Luxury, 14*(3), 45–53.
26. Brown, K., & Patel, S. (2021). Stella McCartney: Pioneering sustainability in luxury fashion. *Green Fashion Insights, 10*(4), 87–95.
27. Miller, J., & Anders, C. (2020). Artisanal approaches to waste management: The petit H model. *Luxury Craftsmanship Journal, 8*(2), 35–42.
28. Chen, Z., Lee, Y., & Park, S. (2021). Minimalism and sustainability in Japanese luxury fashion. *Asian Journal of Fashion Sustainability, 12*(5), 112–120.
29. Lee, H., & Park, K. (2022). Patagonia's worn wear: A case for circular fashion. *Sustainable Textile Review, 15*(1), 9–18.
30. Anderson, M., & Carter, L. (2022). Challenges in adopting sustainable materials in luxury textiles. *Journal of Textile Economics and Policy, 19*(3), 65–72.
31. Garcia, A., & Singh, R. (2021). Technological barriers in recycling high-value textiles. *Textile Science Journal, 11*(2), 76–83.
32. Johnson, E., & Nguyen, M. (2020). Consumer preferences for sustainable luxury products. *Journal of Fashion Marketing and Management, 24*(3), 210–225.
33. Taylor, B., & Smith, R. (2021). NGO collaborations in promoting circular economies. *Fashion Sustainability and Policy Review, 9*(4), 29–38.
34. Anderson, L., Miller, J., & Carter, T. (2023). Regenerative design in luxury textiles. *Innovations in Sustainable Fashion, 21*(2), 23–31.
35. Chen, Y., & Liu, F. (2022). Circular supply chains: A game-changer for luxury textiles. *Textile Research and Innovation Review, 14*(1), 45–52.
36. Lee, H., & Carter, M. (2021). AI applications in textile waste management. *Journal of Advanced Textile Technologies, 9*(4), 112–121.

Pollution Abatement in the Luxury Textile Sector

Benitta Christy P. and Kavitha S.

1 Introduction

Though the luxury textile industry is well known for generating exquisite fabrics and clothing, it also ominously contributes to the adulteration of environment [1]. This manufacturing unit uses a lot of water, current (energy), and chemicals, which is wicked for the environment. Chemicals discharged into water bodies during the dyeing, printing, and finishing processes contaminate them. Emissions and improper waste disposal lead to the contamination of the air and soil. In response, there is a growing demand for luxury firms to adopt sustainable production methods, particularly to address the environmental distresses without compromising the caliber of their goods [2]. With the improved consciousness of natural sustainability, the textile industry has a higher accountability to reduce pollution. Eco-friendly performances and innovative technologies are ahead in popularity as a means of plummeting pollution [3]. To outline the future of sustainable luxury textiles, this chapter examines different kinds of pollution that are formed during the textile industrial process, determines its sources, and investigates both conventional and cutting-edge methods of pollution abatement [4].

Characteristics, including value, distinctiveness, utility, dependability, and durability, are all included in the purposeful dimension of luxury values. Luxury possessions are more uncommon than quick fashion, necessitate dedicated skills, take

Benitta Christy P.
Department of Costume Design and Fashion, PSG College of Arts & Science, Coimbatore, Tamil Nadu, India

Kavitha S. (✉)
Department of Home Science, Mother Teresa Women's University, Kodaikanal, Tamil Nadu, India

Sadhna et al. (eds.), *Crafting Sustainability in Luxury Textiles for a Zero-Waste Future*, Sustainable Textiles: Production, Processing, Manufacturing & Chemistry, https://doi.org/10.1007/978-3-031-86773-6_7

longer to yield, and can use natural raw materials and handmade customs. Thus, luxury and sustainability are connected, as seen by luxury brands' desire for greater quality and longevity [5].

The eco-design method takes into account a product's effects on the atmosphere at every stage of its life cycle, from raw materials to production and usage to the end of its valuable life [6].

A product's resilience to environmental changes in both intrinsic and extrinsic aspects is referred to as its durability. Whereas extrinsic denotes to outside changes that disturb consumer predilections, intrinsic refers to material corrosion and wear and tear. Conversely, resilience refers to the relationship between a product and its surroundings that allows it to recover from significant harm on both an internal and external level [7]. The luxury business trusts heavily on communications for branding and marketing. To create a sustainable brand image, the luxury firms must accentuate the social and conservational features in their marketing. The use of environmentally conscious imagery and videos in media is a usual strategy when selling sustainable features of business. The third option is to disclose societal and environmental features of firms in their company intelligence and websites in accordance with international norms [8].

2 Types of Pollution in Textile Manufacturing

Textile manufacturing is one of the primary sources of pollution. It leads to water, air, and soil pollution. The harmful chemicals and dyes released into water bodies lead to water pollution. The volatile organic compounds released during dyeing and finishing cause air pollution. Textile waste and synthetic fibers also contribute to soil and land pollution, harming the environment.

2.1 Water Pollution: Impact of Dyeing and Finishing Processes

Water pollution often arises from the dye and finishing techniques employed in textile manufacturing processes. During dye production, in particular a significant volume of water is utilized along with substances such as metals that ultimately find their way into water sources, leading to harm to aquatic environments. Also, 200 liters of water are required for every kilogram of fabric to dye it [9] and a considerable portion of this wastewater is containing substances like alkali salts and azo dyes that pose a serious threat to the cleanliness of rivers and lakes [10].

2.2 Air Pollution: Emissions from Textile Factories

There are many pollutants emitted from the textile factories into the atmosphere (VOCs), the carbon dioxide, these sulfur include dioxide, volatile and organic nitrogen compounds oxides. These emissions are mainly the outcome of energy consumption in processes like spinning, weaving, and dyeing, which are energy intensive and mainly based on fossil fuel, especially coal. The emissions of these gases lead to global warming and acid rain, thus worsening the climate change effects [3].

2.3 Soil Pollution: Disposal of Chemical Waste and Its Effects

In the textile industry, soil pollution is due to the incorrect management of solid chemicals discarded. Dyes, plastics, and solvents that are employed in the manufacture of textiles and other related processes can leach into the ground and thus have adverse effects on the water table as well as the environment. Repeated use of these chemicals poses a threat of making the soil infertile and hence unfit for cultivation in the long run [11].

3 Sources of Pollution

The sources of pollution from the textile industry include toxic chemicals during dyeing and finishing processes, untreated factory effluent, and synthetic fibers. Significant sources include the amount of energy and carbon produced by machines and waste created from the speed of production and disposal in fast fashion cycles.

3.1 Chemical Use: Types of Dyeing, Printing, and Finishing Chemicals

The chemicals that are utilized in dyeing and finishing processes are the leading cause of pollution in the textile industry. Here, synthetic dyes, formaldehyde-based resins, and other chemical additives are used to impart the required fabric properties. Most of these chemicals, specifically azo dyes, are hazardous and cancer causing, thus having adverse effects on the health of workers and consumers [12].

Chemical runoff from these processes frequently enters water bodies, endangering aquatic life and ecosystems.

3.2 Energy Consumption: Impact of Energy Use on Pollution

The textile sector is significantly dependent on energy, with a particular emphasis on nonrenewable assets such as coal, and natural gas. The energy requirements throughout different phases of production, including spinning, weaving, and processing, lead to considerable carbon emissions. This energy usage not only exacerbates air pollution but also amplifies the carbon footprint associated with textile fabrication, thereby heightening apprehensions regarding the industry's role in contributing to global warming [13].

3.3 Water Usage

Luxury textile production is notorious for its high water consumption, particularly during the dyeing and finishing stages. Water is required at almost every stage of the manufacturing process, with some luxury textiles, such as silk, necessitating even more water-intensive treatments. Excessive water use has resulted in the depletion of freshwater resources in areas where water scarcity is already an issue, exacerbating environmental and social problems [14].

4 Current Pollution Abatement Techniques

Several pollution-reduction techniques have been implemented in recent years to address the environmental concerns raised by textile manufacturing [15]. Water treatment technologies, such as effluent treatment plants (ETPs), advanced oxidation processes (AOPs), adsorption, coagulation, and flocculation, membrane filtration, and zero-liquid discharge (ZLD) systems [16], are becoming more prevalent for the purification of wastewater prior to its release into the environment. Additionally, particulate emissions from industrial facilities are managed through air filtration systems, including scrubbers and electrostatic precipitators [17]. Nevertheless, these solutions can be expensive and may pose challenges for smaller manufacturers.

Algae, simple yet remarkably adaptable organisms prevalent in aquatic ecosystems, are significantly influencing the textile sector. Companies such as Algaeing have led the way in utilizing algae to create nontoxic, biodegradable fibers, and dyes for textiles [18]. This groundbreaking approach not only delivers an environmentally friendly alternative to traditional materials but also consumes considerably less water and energy, tackling some of the most pressing environmental concerns linked to textile production [19]. Algaeing's process demands transforming algae into a liquid solution that aids as a dye, and by incorporating cellulose, it can be transformed into fibers appropriate for fabric manufacturing.

The textile industry boasts an extensive array of natural materials, each offering distinct properties and advantages. Utilizing these materials transcends mere sustainability; it celebrates the rich textures, unique qualities, and narratives they contribute to our clothing. From the elegant flow of silk to the earthy appeal of hemp, each fabric possesses its own special traits [20]. These materials serve as a reminder that fashion can harmoniously blend beauty with environmental consciousness. As the industry endures to innovate, we can expect to see even more exhilarating developments in natural textiles, assertive the limitations of whatever's possible and prominent the way toward a more ecologically friendly future.

5 Innovative Approaches and Technologies

The luxury textile industry is adopting more innovative pollution control technologies and promoting sustainability. One such advancement is the usage of bio-based alternatives to traditional materials, which are being developed from surprising sources like food waste and ocean matter. For instance, companies such as Keel Labs and TomTex are creating materials from algae and shrimp shells, respectively, that not only offer a sustainable option but also provide the luxurious feel and durability expected from high-end textiles.

Waterless dyeing with supercritical carbon dioxide (CO_2) eliminates the necessity for water and significantly reduces pollution [21]. Digital printing significantly reduces the reliance on chemical dyes and water consumption. Luxury brands are increasingly implementing energy-efficient manufacturing techniques, including the utilization of solar energy and other renewable resources, to minimize greenhouse gas emissions [22]. Additionally, there is a growing trend toward the creation of biodegradable textiles and sustainable fibers, such as organic cotton and hemp, which have a reduced environmental footprint during both production and disposal [4]. The textile sector is also investigating the potential of biodegradable plastics that can safely decompose in natural settings, thereby mitigating the issue of plastic waste.

Beyond material advancements, there is a strong emphasis on enhancing waste management through improved recycling initiatives and zero-waste strategies. These efforts focus on effectively managing various types of plastic waste, incorporating advanced sorting technologies and chemical recycling processes. Furthermore, innovations in textile dyeing and patterning, including enzyme and laser processing techniques, offer promising alternatives to traditional methods by decreasing the use of chemicals, water, and energy, while also minimizing waste generation. The integration of technology is further highlighted by initiatives aimed at increasing supply chain transparency. For instance, blockchain technology is being utilized to monitor the life cycle of products and ensure adherence to sustainable practices throughout production. Additionally, the deployment of microplastic capture devices in washing machines and stormwater systems represents a technological solution designed to prevent these pollutants from contaminating water sources.

These groundbreaking methods not only promote environmental sustainability but also resonate with the luxury industry's dedication to quality and exclusivity. By embracing sustainable technologies, luxury textile producers are establishing a new benchmark for the sector, showcasing that luxury can coexist with environmental stewardship. As these technologies advance and expand, they offer the potential for a more sustainable future in fashion, where luxury textiles play a constructive role in benefiting both the environment and society.

6 Brands That Focus on Sustainable Materials

The fashion industry is witnessing a significant shift toward sustainability [23], and this is particularly evident in the realm of sportswear, active wear, swim suits, lingerie, and underwear. In the realm of sustainable fashion, numerous brands have emerged as frontrunners, championing the use of natural materials and ethical production practices. Several luxury brands are making significant strides in pollution abatement and sustainability. The belief of superior brands are relatively old is a valid talk. True luxury brands, such as Hermes and Louis Vuitton, have histories of over 175 and 160 years, respectively. Ready-to-wear luxury labels, such as Tommy Hilfiger, are relatively new (approximately 30 years old). This exemplifies that the artesian luxury brands that supply actual luxury products are significantly more mature than the brands that provide ready-to-wear luxury [24]. Here are a few leading examples.

Stella McCartney has been a pioneer in sustainable fashion since its inception. The brand avoids using animal leather, feathers, fur, or skins and focuses on cruelty-free alternative materials, process innovation, and circular practices. Patagonia stands out for its commitment to environmental activism and the use of recycled materials in its clothing lines. Known for its transparency, the company has been a pioneer in the sustainable apparel industry for decades.

Gucci, under the Kering Group, has implemented a comprehensive sustainability strategy. This includes using eco-friendly materials, reducing carbon emissions, and ensuring responsible sourcing. Gucci aims to achieve a 50% reduction in its environmental footprint by 2025. Prada has launched several initiatives to promote sustainability, including using recycled nylon and reducing greenhouse gas emissions. The brand is also part of The Fashion Pact, which aims to achieve net-zero carbon emissions by 2050 [25].

7 Identifying Natural Materials in Clothing Labels

Clothing labels serve as a treasured resource, offering detailed information about fabric composition, care guidelines, and the sustainability of the item. To find natural materials, consumers should search for labels that specify the types of fibers castoff. These labels are commonly located inside the neckline, along the side seams, or within the waistband of the clothing.

The fabric composition is generally presented as a percentage, especially when the garment consists of a fiber blend. For instance, a label stating "70% silk, 30% linen" specifies a combination of natural fibers. It is crucial to understand the distinctions between synthetic and natural fibers. When purchasing clothing, it is advantageous to investigate brands recognized for their use of natural materials and to seek out certifications that reflect sustainable practices. Certifications like the Global Organic Textile Standard (GOTS) or OEKO-TEX provide confidence that the textiles adhere to specific environmental and health criteria. By familiarizing themselves with these labels and certifications, consumers can make educated choices that resonate with their values and support a more sustainable fashion industry.

8 Conclusion

Globalization, along with the emergence of new entertainment avenues, has led to an increase in luxury fashion consumers. Concurrently, the rising awareness of social and environmental issues among these consumers has placed significant pressure on companies to integrate social responsibility into their business models. Many consumers perceive that luxury brands embracing sustainable practices are primarily motivated by commercial interests [26]. It is essential to transition from a "fast fashion" paradigm to one of "slow fashion," where garments and accessories are crafted for longevity rather than being discarded after a single season. The essence of slow fashion inherently embodies luxury [27]. We recommend that brands undertake a thorough analysis of the necessity to educate consumers about the sustainable practices they implement, while also conducting research to identify various customer segments interested in sustainable products. In summary, sustainable luxury is closely linked to the manufacturing practices of companies and the behaviors of consumers.

References

1. Berg, A., Magnus, K.-H., Kappelmark, S., Granskog, A., Lee, L., Sawers, C., & Polgampola, P. (2020). *Fashion on climate*. McKinsey & Company and Global Fashion Agenda.
2. Bick, R., Halsey, E., & Ekenga, C. C. (2018). The global environmental injustice of fast fashion. *Environmental Health, 17*(1), 92.

3. Blackburn, R. S. (Ed.). (2009). *Sustainable textiles: Life cycle and environmental impact.* Woodhead Publishing.
4. Choudhury, A. K. R. (2014). Sustainable chemical technologies for textile production. *Textile Progress, 46*(1), 1–75.
5. Hennigs, N., Wiedmann, K.-P., Klarmann, C., & Behrens, S. (2013). Sustainability as part of the luxury essence: Delivering value through social and environmental excellence. *Journal of Corporate Citizenship, 2013*, 25–35.
6. Niinimäki, K., & Hassi, L. (2011). Emerging design strategies in sustainable production and consumption of textiles and clothing. *Journal of Cleaner Production, 19*(16), 1876–1883. https://doi.org/10.1016/j.jclepro.2011.04.020
7. Vanackera, H., Lemieuxa, A.-A., & Bonnierb, S. (2022). Different dimensions of durability in the luxury fashion industry: An analysis framework. *Journal of Cleaner Production, 377*, 134179. https://doi.org/10.1016/j.jclepro.2022.134179
8. Perry, P., & Towers, N. (2013). Conceptual framework development: CSR implementation in fashion supply chains. *International Journal of Physical Distribution & Logistics Management, 43*(5/6), 478–500.
9. Nagurney, A., & Yu, M. (2012). Sustainable fashion supply chain management under oligopolistic competition and brand differentiation. *International Journal of Production Economics, 135*, 532–540.
10. Kant, R. (2012). Textile dyeing industry: An environmental hazard. *Natural Science, 4*(1), 22–26.
11. Singha, K., Pandit, P., Maity, S., & Sharma, S. R. (2021). Harmful environmental effects for textile chemical dyeing practice. In *Green chemistry for sustainable textiles* (pp. 153–164). Woodhead Publishing.
12. Ma, H., Burger, C., Hsiao, B. S., & Chu, B. (2011). Fabrication and characterization of cellulose nanofibers-based thin-film nanofibrous composite membranes. *Journal of Membrane Science, 378*(1–2), 1–8.
13. Karthik, T., & Gopalakrishnan, D. (2014). Environmental analysis of textile value chain: A case study on cotton knitted fabrics. *International Journal of Textile Science, 3*(2), 31–37.
14. Chapagain, A. K., Hoekstra, A. Y., Savenije, H. H., & Gautam, R. (2006). The water footprint of cotton consumption: An assessment of the impact of worldwide consumption of cotton products on the water resources. *Ecological Economics, 60*(1), 186–203.
15. Islam, M. M., Perry, P., & Gill, S. (2021). Mapping environmentally sustainable practices in textiles, apparel and fashion industries: A systematic literature review. *Journal of Fashion Marketing and Management, 25*(2), 331–353. https://doi.org/10.1108/JFMM-07-2020-0130
16. Kallawar, G. A., & Bhanvase, B. A. (2024). A review on existing and emerging approaches for textile wastewater treatments: Challenges and future perspectives. *Environmental Science and Pollution Research, 31*, 1748–1789.
17. Mohan, S., Oke, N., & Gokul, D. (2021). Conventional and zero liquid discharge treatment plants for textile wastewater through the lens of carbon footprint analysis. *Journal of Water and Climate Change, 12*(5), 1392–1403.
18. https://edition.cnn.com/style/article/israel-algae-environment-textiles-spc-intl-hnk/index.html
19. Yadav, G., et al. (2021). Mechanism and challenges behind algae as a wastewater treatment choice for bioenergy production and beyond. *Fuel, 285*, 119093. https://doi.org/10.1016/j.fuel.2020.119093
20. Raja, T., & D, Y. (2023). Analysis of the properties of a hemp fibre derived from Cannabis sativa as a renewable source. *Biomass Conversion and Biorefinery, 13*, 15149–15159. https://doi.org/10.1007/s13399-023-04861-y
21. Eren, H. A., Yiğit, İ., Eren, S., & Avinc, O. (2020). Sustainable textile processing with zero water utilization using super critical carbon dioxide technology. In *Sustainability in the textile and apparel industries: Production process sustainability* (pp. 179–196). Springer.

22. Parisi, M. L., Fatarella, E., Spinelli, D., & Basosi, R. (2014). Environmental impact assessment of an eco-efficient production for coloured textiles. *Journal of Cleaner Production, 108*, 514–524.
23. Cervellon, M. C., & Shammas, L. (2013). The value of sustainable luxury in mature markets a customer-based approach. *Journal of Corporate Citizenship, 52*(12), 90–101.
24. Rahman, S., & Yadlapalli, A. (2015). Sustainable practices in luxury apparel industry. In M. A. Gardetti & S. S. Muthu (Eds.), *Handbook of sustainable luxury textiles and fashion, environmental footprints and eco-design of products and processes*. Springer Science + Business Media. https://doi.org/10.1007/978-981-287-633-1_8
25. https://sustainabilitymag.com/sustainability/top-10-sustainability-strategies-of-luxury-fashion-brands
26. Achabou, M. A., & Dekhili, S. (2013). Luxury and sustainable development: Is there a match? *Journal of Business Research, 66*(2013), 1896–1903.
27. Olorenshaw, R. (2011). Luxury and the recent economic crisis. *Vie & sciences de l entreprise, 188*, 72–90.

Recycled Elegance: Evaluation of Recycling in Luxury Textile Manufacturing

Jayapriya S. and Suresh S. N.

1 Introduction to Luxury Textiles

The luxury textile segment is a fast-moving industry that sells to an advanced consumer group whose desire for high quality is outweighed only by the need for creative, sustainable, and innovative textile design and manufacture. Luxury textiles are differentiated by high-quality raw materials, elaborate designs, and high-end production processes that create a superior final product. This industry is characterized by high-status appealing brands, giving weight to the point of small production runs and limited-edition releases. Luxury textiles are tested to ensure they meet both industry standards and consumer expectations. Testing is vital for durability, colorfastness, feel, environmental impact, and resistance to higher and lower wear from uses. Each textile should be verified and tested that it proves its standard [1].

Sustainable practices have also caused a shift in the demands placed on luxury textile testing. As consumers become more concerned with sustainability, they are seeking textiles that feel more like their favorite eco-friendly footwear and create less of a carbon footprint. Your test results now include environmental metrics like the carbon footprint and biodegradability of materials. Considering just the fact that ethical dynamic of test has been included, the testing criteria for luxury textile testing have become more extreme. This chapter will present the main luxury textiles testing methods. It explores traditional and new techniques from sustainability testing to digital technologies and their role in the industry. This discussion will shed

Jayapriya S. (✉)
Associate Professor and Head, Department of Costume Design and Fashion, Nehru Arts and Science College, Coimbatore, Tamil Nadu, India

Suresh S. N.
Associate Professor and Dean, School of Science, Rathinam College of Arts and Science, Coimbatore, Tamil Nadu, India

Sadhna et al. (eds.), *Crafting Sustainability in Luxury Textiles for a Zero-Waste Future*, Sustainable Textiles: Production, Processing, Manufacturing & Chemistry, https://doi.org/10.1007/978-3-031-86773-6_8

light on how luxury textile testing can be customized to the long-running traditions of quality and the evolving requirements of the contemporary consumer [2].

2 Luxury Textiles

Luxury textiles possess good quality, feature, and paucity, uniting sporadic extraordinary fiber with methodological proficiency of skilled craftsperson. They can require labor-intensive production processes with the finest of fibers, such as silk and vicuña wool. While being some of the softest materials around, many luxury textiles are also very durable and come at a far higher price point due to their luxurious appearance. They are characterized by the materials processed and tools used in their manufacture as well as with limited-run production. They are made from high-quality fibers such as silk or exclusive animal wools (cashmere and others), with meticulous production processes. Many fabrics are produced from handwoven techniques or with modern technology so that it gives a better texture, toughness, as well as make the aesthetics reach its peak. They are luxury in as much as they tend to be highly exclusive, rare, or sourced ethically and sustainably, which does not apply on a mass scale for regular textiles [3].

Luxury textiles in fashion and interior decoration are given below with their historical context as ancestors to fabric by the yard. Since ancient civilization, luxury textiles have been reserved for the elites of societies in places like Egypt, Rome, and China, where only royalty had access to silk, which at that time was a rare fabric. Luxury textiles have influenced the couture and Haute Couture collections of fashion houses and were destined for royal palaces or luxury homes, namely velvet fabrics and damask. Modern-day luxury markets continue to covet these textiles [4].

3 Fabrics Utilized in Luxury Textile Production

Silk is an incredibly old and rare luxury textile, with its earliest production going back to Ancient China around 2700 BCE. Extracted from the cocoon of silkworms in a process called sericulture, silk is produced by very cautiously extracting the strands. The lustrous sheen, softness, and lightweight strength of silk has made it a highly sought-after material. It is highly absorbent of moisture and has a wonderful drape, perfect for luxury fashion applications such as upholstery fabrics or accessories [5].

Cashmere is an extremely desirable fiber that comes from the downy undercoat of cashmere goats raised in Mongolia and Tibet. It feels luxuriously soft, is lighter in weight, and also offers better insulation than sheep wool. Cashmere is known as a luxury item because there are only so many goats in the world, and getting the cashmere requires a lot of work. These are often used in luxury knitwear, scarves, and blankets [6].

Vicuna wool is the fiber from the wild vicuñas in the Andes, which is one of the rarest, most luxurious, and expensive fibers in the world. It can be shaved every 3 years, thus the fiber becomes rarer. While alpaca wool is easier to find than vicuna, it is similarly priced for its softness, lightweight warmth, and hypoallergenic quality. Both are used in the luxury textile production of haute fashion and interior decoration, being both rare and desirable, due to the high quality [7].

Historically, high-end fashion has been linked with exotic leathers (crocodile, ostrich, python) and fur, including mink and fox. But with growing concerns about animal welfare, these materials are facing more and more scrutiny. Luxury brands are increasingly embracing ethical alternatives from plant-based leather to lab-grown fur to cruelty-free materials. Sustainable luxury is the direction the industry is heading, offering not just a moonshot solution but high-quality alternatives as well [8].

4 Luxury Textile Recycling

Luxury textile recycling is a rising preparation intended to drop the environmental influence of the fashion industry by recycling first-class materials. This procedure includes the gathering, categorization, and reprocessing of luxury textiles, such as silk, wool, and cashmere, to generate novel fashions or goods. Though the excellence of these reprocessed fibers is critical to uphold the best status of luxury trademarks, progressions in reprocessing skills, such as biological and powered recycling, are refining the robustness and aesthetic potentials of recycled textiles. Foremost brands like Stella McCartney then Gucci have initiated the mixing of recycled fibers into their assortments, representing the possibility for sustainability in luxury technique. Though experiments continue in topping processes whereas safeguarding that recycled materials encounter the rigorous needs of high-end consumers [9].

The luxury textile sector has come under fire for its use of scarce, high-end materials. Rethinking recycling: 1% of recycled fabric. We all know luxury goods produce mountains of waste, from fabric scraps and resellers, to mattresses stuffed with unsold stock. With the increase in environmental awareness, recycling of luxury textiles is necessary to save the indirect house using luxury fabrics and decrease landfill waste. Such a transition helps a more extensive circular economy, where resources are recycled instead of going to waste [10]. The multilayered nature of luxury fabrics and their workmanship can make recycling difficult. For example, many luxury fabrics such as silk, cashmere, and blended fibers are almost impossible to decompose without harming their intrinsic characteristics. For one, the added details, embellishments, and unique branding that luxury items are known for also make them less able to be recycled without losing their original appeal. It will require creativity for the industry to find solutions to mitigate these challenges while retaining the aspirational nature of luxury [11].

Global textile recycling is now taking momentum with increasing importance of lowering textile rubbish and recollecting the good materials. Higher-end products are made from recycled input via the upcycling method, while breakthroughs in mechanical and chemical recycling techniques enable the recycling of complicated textiles. Consumer interest in sustainable fashion has seen a boom in recent years, causing an increasing number of luxury brands to adopt recycling programs [12].

5 Assessing Recycled Luxury Fibers

Measuring recycled luxury fibers includes appraising both their environmental impact and presentation in luxury textile manufacture. Main factors include the toughness, feel, and aesthetic makings of recycled fibers, confirming that they encounter the great standards predictable in premium goods. Furthermore, the impression of recycling on physical possessions such as springiness, strength, and color retention is vital [13].

Recycling luxury sector resources has one major challenge, which is quality insurance by necessity virgin fibers should have more durability as well as look and feel than recycled fibers. Fibers that get recycled in recycling factories are generally shorter because of the Schermer method, so especially those gained from mechanical recycling are shorter and less elastic compared to virgin fibers. But the increasingly strict specifications required of luxury items can only be met if new recycling steps continue to boost the quality of materials. They still need to go through a lot of tests to check if this will work over a long period of time [14].

Figure 1 likely summarizes the phases of luxury textile manufacture and testing, from raw material obtaining to the finishing quality valuations. It establishes the complicated developments tangled in checking required standards, such as material testing, sturdiness checks, and sensory assessments, crucial for upholding luxury position in textiles.

5.1 Cost of Recycling Luxury Textiles and Their Implications on Price

In fact, recycling luxury textiles usually costs more than employing virgin materials because the processes involved are more complex. Some techniques do require specialized equipment (like those for mechanical recycling), and others like chemical recycling and upcycling come with high costs such as the demand for skilled labor and innovation. While some consumers will pay a higher price for sustainable goods, these costs can push the price of recycled luxury products higher. There is still the other side of sustainability, that is, economic viability or profitable growth, which is equally important for brands to consider to making recycling a commercially viable part of their business model.

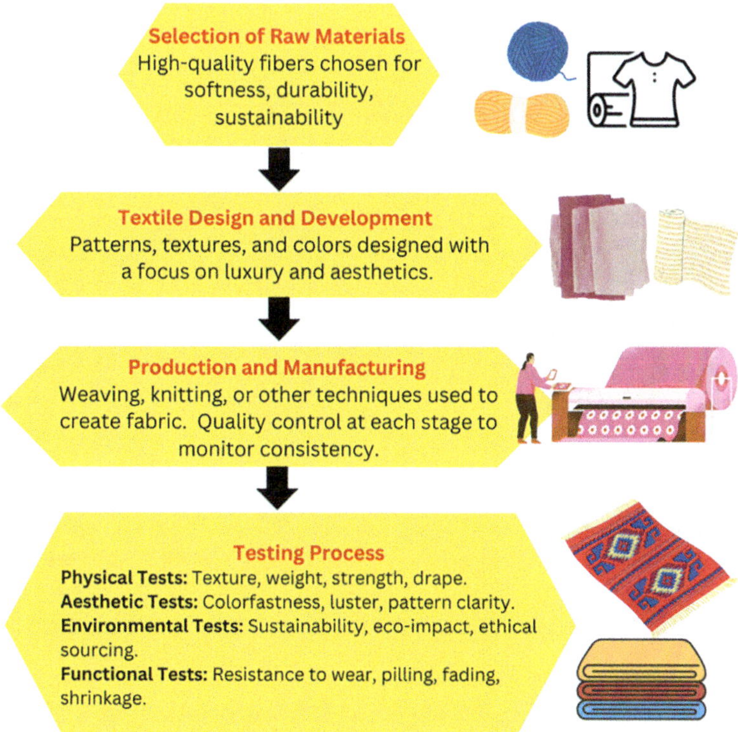

Fig. 1 Luxury textile production and testing process [15]

To analyze the environmental benefits of recycling luxury textiles and reducing the number of new textiles produced, compare the two by using a method called life cycle assessment (LCA). It is through life cycle assessment (LCA) that the advantage of recycled textiles over new textiles can be assessed in terms of environmental impact. Life cycle assessment (LCA) takes into account energy and water use and carbon emissions throughout all phases of production. Key insights through finding recycled luxury textiles yield lesser environmental footprint compared to a new textile as it cut down on resource extraction and waste. Well-executed LCAs help brands understand the value of recycling and understand how to allocate dollars within their sustainability strategy [16].

Figure 2 illustrates the various manufacturing approaches done in apparel manufacture, showcasing the main processes such as preparation of fabric, its cutting, stitching process, and finally finishing. It also highlights the phases involved in converting raw materials into a final finished garment, confirming functionality, aesthetics, and quality.

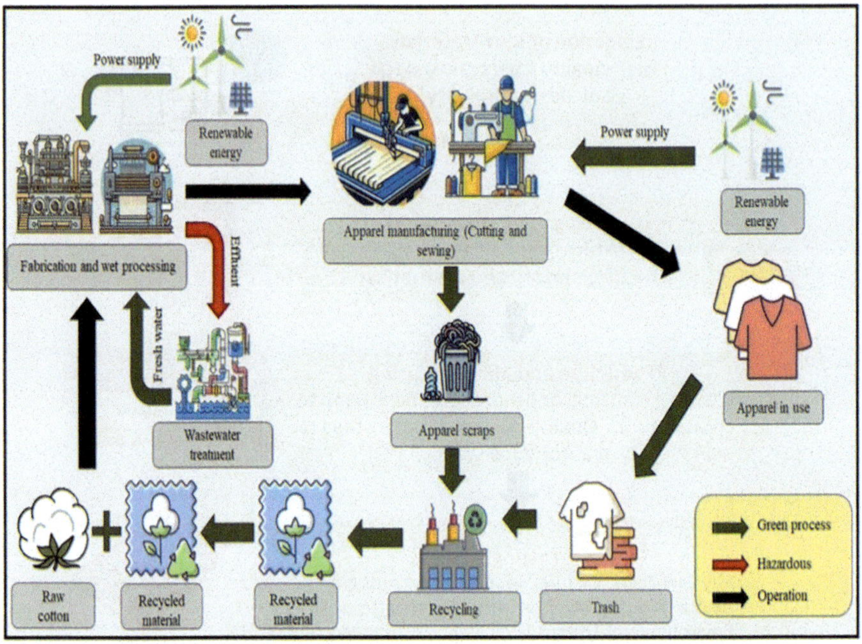

Fig. 2 Manufacturing method for apparel [17]

6 Testing Methods of Recycled Luxury Textiles

In addition, assessing the luxury textiles through rigorous testing is necessary to ensure that recycled fibers uphold the same quality as virgin fibers. Textile testing methods play a significant role during evaluation [18].

(a) **Evaluation of strength and durability**: The fibers that are recycled, specifically by processing them mechanically, may have some integrity loss compared to virgin fibers. Tensile strength testing assesses how much force it takes to break or tear the fabric to confirm that recycled textiles have sufficient strength for high-end use without sacrificing performance. Durability tests such as abrasion resistance, which mimic years of use, confirm that luxury can be sustainable as recycled textiles hold their luxurious touch to the skin after extended use [19].
(b) **Fiber length and fineness testing:** Fine fibers like cashmere silk and wool are used to make luxury fabrics. However, certain aspects like the softness and texture of the material are affected due to mechanical recycling shortening the length of fibers. Tests for fiber length indicate how much of original quality is remaining and fineness testing makes sure that recycled fibers stay as shimmery, smooth, and soft as luxurious textiles [20].
(c) **Colorfastness testing:** One of the hallmarks of luxury textiles are their bold, deep colors. Colorfastness testing subjects fabrics to conditions, including

washing, exposure to light, and rubbing to determine the resistance of the dyes to the recycled fibers. High-end luxury fabrics used in the fashion field are at present designed to have the same aesthetic characteristics as virgin textiles so that recycled textiles are aesthetic never less than virgin ones [21].

(d) **Pilling resistance testing:** Pilling, fiber balls of any size and shape on the surface of fabric, can ruin the luxe aesthetic of many textiles. Textiles that have been mechanically recycled are particularly susceptible to pilling. Pilling resistant tests like Martindale tester, which is a well-known abrasion resistance test, without the use of the pill tester will only determine the surface-driven wear of the fabric [22].

(e) **Softness and tactile quality assessment:** Softness, stiffness, and elasticity of recycled textiles were measured by subjective hand feel tests, and through objective measurements using system such as the Kawabata Evaluation System (KES). This allows all recycled fabrics to have their signature look and luxury feel [23].

(f) **Moisture absorption and thermal regulation testing:** Moisture-wicking and thermal regulation properties of luxury textiles like wool and cashmere. The testing of the attributes ensures that the recycled materials are still able to perform on the same scale, providing comfort in varying conditions [24].

(g) **Shrinkage and dimensional stability testing:** Luxury textiles should hold their shape and dimensions over a period of time. Shrinkage testing measures the changes in recycled textiles that occur after washing, while dimensional stability tests determine if the fabric will sag, grow, or shrink. Because recycled luxury textiles can bring longer sustenance, dimension stability is essential [25].

(h) **Environmental testing:** Chemicals used in textile production have become more conscious of luxury brands. Testing recycled textiles for harmful chemicals ensures that recycled processes do not introduce toxic substances. OEKO-TEX Standard 100 and Global Recycled Standard help verify the safety and eco-friendliness of recycled luxury textiles [26].

6.1 Subjective Assessment

(a) **Texture:** The texture of fabric is vital to its visual and tactile appeal. It can vary from smooth to rough, soft to crisp, and everything in between, each variation adding unique qualities to the fabric. Texture not only enhances the sensory experience but also impacts perceptions of quality and luxury. Moreover, texture influences functionality, affecting a garment's drape and movement [27].

(b) **Color:** Color is the most obvious and meaningful characteristic of textiles and often evokes the strongest emotional responses in the first 2–3 s of viewing. Key measures for quality are the richness, depth, and uniformity of the color, and its fade resistance. Since market trends are crucial in designing because market trends also speak about color, colorfastness becomes one of the primary tests one must conduct to ensure quality [28].

(c) **Appearance:** Appearance of the gloss delivers the complete visual effect of a textile such as color and texture, but also luster, pattern, and clarity. It is how these elements mix together to produce a beautiful end result. The visual or aesthetic quality of a textile sends visual messages about luxury, comfort, and even functionality that affects both perceived and actual value. Evaluation of aesthetic qualities in addition to physical and chemical property measurement is an equally important phase in textile quality testing. This part of the testing has to do with expertise in things like fashion and design trends, consumer tastes, and cultural movements. The technical textile knowledge here takes the dive into creative designer intuition, highlighting the desire for those textiles that are not just resilient, functional, and protective, but also can be crispy and charming [29].

The testing methods, in turn, will guarantee that the luxury character of the textile output will remain unchanged even if these textiles are recycled. Rigorous testing is done to ensure recycled materials meet the high expectations that go with being a luxury good [30].

7 Future Directions

As the luxury textile industry progresses, future investigation must discover the technical innovations that drive and create recycling procedures further efficient and lucrative. Precisely, progressions in biochemical recycling methods, such as those recycled to break down raw materials for recycle, understand that there is a great promise for refining the superiority and variety of reprocessed textiles existing for high-end style. Furthermore, studies on the influence of reprocessing on fiber properties, such as strength, springiness, and color preservation, are essential to guarantee that recycled materials encounter the difficult standards predictable in luxury textiles [31].

Customer insights of reused luxury goods continue a critical factor in their widespread acceptance. The investigation should examine the psychological barriers that stay clients after approval of recycled products, predominantly in the luxury section. By knowing the factors that affect buying choices—such as trademark loyalty, ecological values, and the awareness of recycled materials as lesser—luxury brands can adapt their advertising approaches to echo with their target viewers. Reviews and focus groups might offer appreciated feedbacks as like in what way luxury customers opine sustainability and whether the professed value of recycled products bring into line with the brand's appearance [32]. Ultimately, the future of recycled luxury textiles will depend on a combination of technological innovation, consumer education, and ethical production practices. By continuing to push the boundaries of sustainable textile manufacturing, the luxury fashion industry can play a leading role in shaping a more eco-conscious future for fashion worldwide [33].

8 Conclusion

Luxury textile testing is an integral part of maintaining the high standard asserted in this exclusive trade. Luxury textiles have undergone immense testing to ensure that only the finest products are delivered to the market, a testament to textile manufacturers and their commitment to only the best. The test processes guarantee that the textiles meet and even exceed the expectations of luxury consumers. Testing standards need to keep pace with industry evolution. Consumer demand for sustainability and ethical transparency has changed the nature of textile testing. Modern sustainable testing protocols assist luxury brands ensuring your goods are in line with modern ideals. That said, excellence has become manifestation of sustainable excellence as a signature of delicateness, sturdiness, and beauty of the fabric.

Testing goes at the center of any evolution of the luxury textile industry, which is certainly undergoing rapid changes. Luxury textile brands can walk the line of tradition and innovation by progressively refining testing methods. For instance, the textiles need to pass extreme tests and thus the industry is a pioneer in sustainability. The finest part about the luxury textile sector is having been committed to quality and sustainable practices, which ensures it has a secure place in the future of fashion.

References

1. Fletcher, K. (2014). *Sustainable fashion and textiles: Design journeys*. Routledge.
2. Gwilt, A. (2015). *Fashion Design for Sustainability*. Bloomsbury.
3. Mont, O., & Heiskanen, E. (2014). Luxury and the circular economy: Creating value through sustainable innovation. *Journal of Sustainable Luxury*.
4. Hiner, S. (2015). *Luxury fashion: A global history*. Oxford University Press.
5. Harris, J. (1999). *5000 years of textiles*. British Museum Press.
6. King, J. (2011). *Weaving: The art and science*. Thames & Hudson.
7. Wild, J. P. (2018). *Textiles in ancient Mediterranean luxury*. Routledge.
8. Beard, N., & Hiller, C. (2016). *Luxury textile processing and finishing*. Elsevier.
9. McKinsey & Company. (2023, December 1). *Scaling textile recycling in Europe—Turning waste into value*. https://www.mckinsey.com/industries/retail/our-insights/scaling-textile-recycling-in-europe-turning-waste-into-value
10. Lewis, T. (2021). *Circular fashion: A sustainable approach to fashion*. Bloomsbury Academic.
11. Webster, K. (2017). *The circular economy: A wealth of flows*. Ellen MacArthur Foundation Publishing.
12. Chapagain, A. (2016). *Textiles and the environment: Environmental impact of the textile supply chain*. Springer.
13. Sustainable Brands. (2017, June 1). *Stella McCartney commissions first-ever LCA for manmade cellulosic fibers*. https://sustainablebrands.com/read/stella-mccartney-commissions-first-ever-lca-for-manmade-cellulosic-fibers
14. Stengel, O. (2017). *Sustainable materialism: Environmental and social sustainability of materials*. Palgrave Macmillan.

15. Bhikari Charan, S., Panda, C., & Mukhopadhyay, S. (2021). Sustainable pretreatments in textile wet processing. *Journal of Cleaner Production, 289*, 125613. https://doi.org/10.1016/j.jclepro.2020.125613
16. Payen, J. (2020). *Innovations in luxury textile recycling: A technical overview*. Textile Technology Press.
17. Rahaman, M. T., Pranta, A. D., Repon, M. R., Ahmed, M. S., & Islam, T. (2024). Green production and consumption of textiles and apparel: Importance, fabrication, challenges, and future prospects. *Journal of Sustainable Fashion & Textile Technology, 10*(1), 35–46. https://doi.org/10.1016/j.jfata.2024.01.010
18. Eurofins. (n.d.). *Fundamental textile testing: Mechanical and physical tests*. https://www.eurofins.com/textile-leather/articles/fundamental-textile-testing-mechanical-and-physical-tests/
19. Centexbel—VKC. (n.d.). *Detecting recycled textile materials in final products*. https://www.centexbel.be/en/testing/detecting-recycled-textile-materials-final-products
20. Niinimäki, K. (2018). *Sustainable fashion in a circular economy*. Routledge.
21. InTouchQuality.(2016,October6).*5colorfastnessteststopreventtextilefadingandstaining*.https://www.intouch-quality.com/blog/5-color-fastness-tests-to-prevent-textile-fading-and-staining
22. Zięba, J., & Frydrych, I. (2020). A method for the assessment of textile pilling tendency using optical coherence tomography. *Materials, 13*(13), 2970. https://doi.org/10.3390/ma13132970
23. Zhou, J., & Xu, B. (2021). Prediction of pilling grade of alkali-treated regenerated cellulose fabric using a fuzzy inference system. *The Journal of the Textile Institute, 112*(10), 1691–1699. https://doi.org/10.1080/00405000.2021.2012909
24. Zhou, J., & Xu, B. (2021). Study on the moisture absorption and thermal properties of hygroscopic exothermic fiber-based materials. *Polymers, 12*(1), 98. https://doi.org/10.3390/polym12010098
25. Zhou, J., & Xu, B. (2021). Thermal management with innovative fibers and textiles. *National Science Review, 11*(10), nwae295. https://doi.org/10.1093/nsr/nwae295
26. Allwood, J. M., & Laursen, S. E. (2008). *Well dressed? The present and future sustainability of clothing and textiles*. University of Cambridge.
27. Li, Z., & Wang, X. (2019). A subjective evaluation method for textile fabric comfort based on sensory testing. *Textile Research Journal, 89*(8), 1562–1571. https://doi.org/10.1177/0040517518771619
28. Kumar, A., & Singh, R. (2018). Sensory evaluation of fabric properties and their impact on consumer preference. *The Journal of the Textile Institute, 109*(12), 1743–1750. https://doi.org/10.1080/00405000.2018.1497867
29. Choudhury, S., & Sharma, S. (2017). Evaluation of tactile comfort properties of fabrics: A subjective and objective analysis. *Fibers and Polymers, 18*(4), 723–731. https://doi.org/10.1007/s12221-017-7735-1
30. Pookulangara, S., & Shephard, A. (2013). Slow fashion: A paradigm shift in luxury consumption. *Fashion Theory Journal*.
31. Bick, R., Halsey, E., & Ekenga, C. C. (2018). The global environmental injustice of fast fashion. *Environmental Health, 17*(1), 92. https://doi.org/10.1186/s12940-018-0433-7
32. H&M Group. (2020). *H&M conscious actions sustainability report 2020*. https://hmgroup.com/sustainability/
33. Shishoo, R. (2017). *Textile processing and properties: Preparation, dyeing, finishing and performance*. Woodhead Publishing. https://doi.org/10.1016/B978-0-08-100660-1.00006-9

Consumer Behavior in Sustainable Luxury Textiles

P. B. Shree Sahana and E Mokanaasri

1 Introduction

The luxury textile sector has been particularly linked with aspects such as quality, elegance, and superior quality material. However, pressure to hold sustainable practices increases as awareness of environmental issues expands across the globe [1]. This movement is caused by the arrival of a new generation of consumers who value luxurious and high-quality goods and services while being environmentally conscious and socially responsible [2]. These are the new-age consumers who are increasingly becoming conscious about the environment, and they are changing the face of luxury business by making them answerable for the kind of fabrics they use, where they get them from, and what new sustainable technology they are bringing into this sector [3].

Luxury and sustainability can be connected to the production of only specific luxury products, which involve the utilization of rare, special, and long-lasting materials refined to endure for generations [4]. It can be stated that sustainability has become not only a special interest regarding luxury fashion brands but is a must for brands and customers' loyalty. Concerns like climate change and the use of natural resources are informing that the place of sustainability as a factor of the fashion industry is indeed critical. Those who do not pay heed to those issues compromise

P. B. Shree Sahana (✉)
Department of Fashion Technology, National Institute of Fashion Technology, Chennai, Tamil Nadu, India

E. Mokanaasri
Department of Fashion Design, School of Arts and Design, Woxsen University, Hyderabad, Telangana, India

Sadhna et al. (eds.), *Crafting Sustainability in Luxury Textiles for a Zero-Waste Future*, Sustainable Textiles: Production, Processing, Manufacturing & Chemistry, https://doi.org/10.1007/978-3-031-86773-6_9

a large chunk of their customer base and become insignificant in an emerging green-conscious world [5].

This chapter covered the key factors shaping the eco-conscious luxury consumer, examining market trends in sustainable luxury fashion and how consumer preferences evolve toward sustainable textiles. It highlights the growing importance of communicating sustainability to luxury consumers, focusing on effective strategies to engage them. The chapter also includes case studies of successful sustainable luxury brands, showcasing how they have integrated eco-friendly practices and communicated their values to the market. These insights provide a comprehensive understanding of the shifting dynamics in the luxury fashion industry toward sustainability.

2 Understanding the Eco-Conscious Luxury Consumer

Understanding the consumer is valuable for the luxury industry, which increasingly incorporates sustainability into its agenda, to gain insights on the values, psychological characteristics, and behaviors of sustainable luxury consumers. According to these enhanced "green" and "ethical" consumers, the market demand is being set that puts pressure on luxury brands to become more sustainable [6]. In addition to the leading role luxury fashion retailers can play in encouraging sustainable consumption, consumer emotion of guilt has also been noted to influence sustainable consumer behavior [7].

2.1 Characteristics of Eco-Conscious Luxury Consumers

These consumers are keen on the environmental implications of consuming certain brands, and therefore, they go for materials that are environmentally friendly in production. Most of their purchasing decisions are influenced by such factors as the desire to minimize their carbon footprint, save natural resources, and protect the biodiversity of the species [8]. By so doing and especially those with such values as Stella McCartney that use biodegradable fabrics as well as employing a no-waste policy, customers are attracted [9]. Sociocultural factors are also a consideration in the development of a condition. Eco-consciousness is very important because consumers consider it a form of culture that embraces a sustainable way of living. This is a group that is sensitive to trends within their circles of influence and to environmentalism. Sustainable luxury brands use this effectively by discussing the ethical values of the products they are marketing, such as artisanal, fair trade, and Indigenous manufacture [10].

Ego-centered motives are achievement and narcissistic motives. Many luxury niche consumers gain personal satisfaction from identifying with brands that offer exclusivity and environmentalism. This is not just about projecting wealth but also about moral mileage or leadership in the environment [11]. Patagonia and other brands delivering activism and ethical manufacturing meet this particular kind of status symbolism [12].

2.2 Psychological and Demographic Profiles

The important factor is that luxury consumers who are concerned about sustainability are more educated and earn more money, and younger consumers like Millennials and Gen Z and also a considerable percentage of baby boomers. These younger consumers who have been brought up in the midst of an awareness campaign regarding environmental issues are much more likely to consider the environmental and social cost of their purchases [13]. The respondents revealed a willingness to pay a premium for their environmentally friendly luxury goods as long as that does not compromise luxury quality and sustainability. This demographic goes for hedonic consumption, which means a proper combination of luxury and ethical values that will meet all the expectations of consumers, which is a challenge for luxury brands [14].

2.3 Behavioral Patterns and Purchase Decisions

Thus, the intention of eco-conscious luxury consumers is more selective in their behavior as consumers. They research well, pay attention to the supply chain, and call the bluff of the brands demanding that they come clean about their sustainability initiatives [15]. Research indicates that these consumers are driven by ethical, distinctive value appeals, but they also seek to validate luxury fashion brands' sustainability initiatives [16]. Figure 1 shows the expected decision-making process of an eco-friendly luxury consumer. It provides a concise guideline, outlining the key steps and factors influencing their purchase decisions, offering valuable insights into their priorities and considerations when selecting sustainable luxury products.

Fig. 1 Consumer decision-making process for sustainable luxury [17]

3 Market Trends in Sustainable Luxury Fashion

Innovatively, the luxury fashion industry is experiencing a shift of paradigm as sustainability becomes part of the bipolar value in fashion. All of these markets are the dynamics that have been established to define the future of luxury textiles with respect to environmentalism and consumerism. Those brands that have included sustainability in their business models and manufactured products are being rewarded by consumers in the environmentally friendly luxury goods market.

3.1 Rise of Sustainable Materials

The use of sustainable materials is one of the era-defining shifts in the luxury fashion industry, today. Modern luxurious brands are shifting toward less conventional materials and fabrics such as biodegradable compostable and recycled fibers [18]. The future of sustainable luxury could very well be grown under a glass Petri dish.

Sustainable materials grown in a laboratory are being developed by innovative designers for use in the production of their work, such as bio-fabricated leather, synthetic silk, and vegan wool. They resemble the conventional types by their nature, though without resulting environmental or ethical effects [19].

Innovations like organic cotton, Tencel, and hemp also obtained preference more than the conventional materials because of the less harm they press on the environment. Other firms including Stella McCartney have also started using recycled polyester and Econyl, a fiber derived from recycled ocean plastics to reduce waste and material use [20]. This move is not a simple question of form, but of content: to make fabrics that reflect the principle of sustainability that now guides the consumer and his/her desire for environmentally friendly fashion. It is for this reason that these consumers expect luxury brands to have environmentally friendly products that are also as attractive as their counterparts as possible. Therefore, luxury brands are trying out advanced eco products that are also expensive and have less impact on the environment [11].

3.2 Circular Fashion Design

The growing trend among these consumers toward purchasing secondhand luxury items is a sustainable alternative. Approximately 23% of wealthy individuals have already bought preowned luxury items, with many more expressing interest in doing so in the future. This shift signifies a growing acceptance of circular fashion principles within the luxury sector, as shown in Fig. 2. Circular fashion, which involves the life cycle of textiles through wear, reuse, and recycling, is steadily gaining popularity as an effective approach to minimizing waste [22].

Sustainable luxury fashion circularity concepts are durability, reclaimability, and reused or recyclable components and materials in circular fashion design. It is in contrast to linear models where products are used for one life and are disposed of, raising the end-of-line concept whereby garments can be upcycled or recycled [23]. Brands engage consumers through education on responsible ownership, encouraging a shift toward quality over quantity. Innovative business models, such as rental and resale services, further support sustainability by reducing waste and resource consumption. Despite challenges like traditional sales focuses and supply chain limitations, circular fashion aims to redefine luxury consumption in alignment with environmental goals [24].

3.3 Technology Integration

The sustainable luxury communication is employing innovations in its application with the wearable technology and blockchain. For instance, wearable technology opens a door for brands to design smart apparel that can evaluate the number of

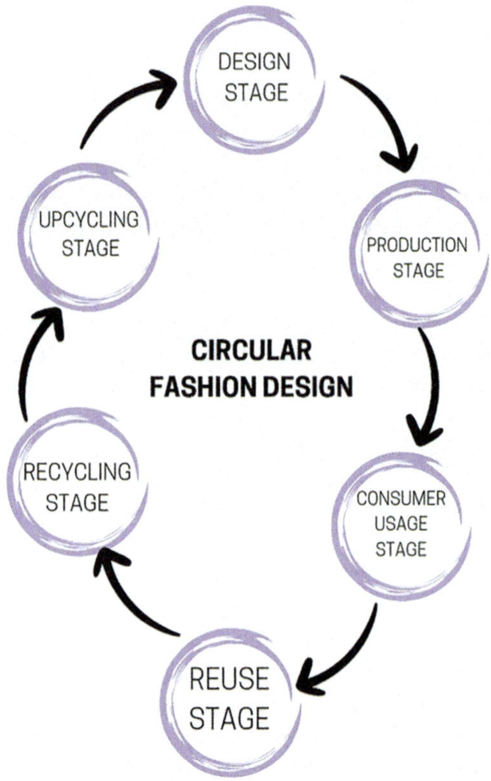

Fig. 2 Circular fashion flowchart [21]

environmental contributions and offer the buyer useful information about the sustainability of the product [25]. This kind of communication experience not only increases the degree of consumers' participation but also stresses the brand environment protection.

The use of blockchain is being done to bring the transparency and accountability into the supply chain. With the rising complications associated with the source of materials in luxury goods, blockchain lets several brands prove that sources of their materials are fair and genuine. This level of traceability has been widely adopted by the eco-conscious consumers who demand the credibility of sustainability-related information [26].

Through television and documentary ads and social media influencers, more and more consumers are being educated on the sustainability of fashion. These platforms explain the damages fashion makes to the environment and encourage the community to embrace more responsible decisions. In documentaries, people are presented with the negative impacts that fast fashion has and the implication for them to go for better options. Relevant experts and influencers in social media, with an interest in sustainable living, work with brands, sharing their valuable perspective and influencing people's decisions regarding environmentally friendly products. This awareness is growing, thus encouraging more consumers to run for environmentally friendly and ethically produced products [27].

4 Consumer Preferences in Sustainable Luxury Textiles

Consumers of sustainable luxury textiles are changing in that many consumers are now concerned with both luxury and sustainability. Over the last few years, there has been quite a significant trend of customers desiring products that are both exclusive and environmentally friendly. Since the consumers are more conscious about what they use, those that meet quality are unique and made with precise value propositions—quality at the core, sustainability, minimal waste, ethical working conditions, and so on are likely to be preferred [28]. COVID-19 has reinforced this trend by focusing on durability and sustainability of purchased goods [29]. Furthermore, consumers themselves are now much more of a segmented audience, and specific segments may care most about waste or supply chain, for example. Such changes in consumers' preferences are gradually affecting the luxury industry, forcing companies to adapt and accommodate new values of environmentally concerned consumers [30].

4.1 Quality and Uniqueness

The consumers continue embracing the traditional physical properties of luxury, which entails quality and exclusivity among the aspects of luxury. In recent years, there has been an increase in the mass market production of luxury goods, specifically in the retail trade, where uniqueness, tailored to customer needs and desires, is promoted as a worldwide freedom of choice trend [31]. Becoming increasingly apparent, this trend in the direction toward more personalized luxury not only increases the utility value in the eyes of the consumer but also serves to build a more profound emotional bond. Luxury brands find that mixing traditional luxury symbols with modern individualization allows consumer needs to be met even concerning luxury goods [32].

4.2 Sustainable Materials

There is a large consumer base who is interested in products that are made from materials that are environmentally suitable or reusable. Such measures of zero-waste working in production were especially popular with buyers sensitive to the impact of their consumption on the environment [33]. This movement is not limited to the type of materials used only but is associated with other sustainable approaches such as energy and also the minimization of wastage together with labor rights. There are higher tendencies shown by the luxury consumers to approach only those products that are environmentally sensitive and have clear visibility of the production process. As pointed out by various publications, today's luxury brands are also paying a lot of attention to the use of renewable energy and the reduction of waste and energy emissions throughout their production cycles. Such a change in the trends is not solely credited to the pressure from the consumers but can significantly be seen as an innovation the luxury brands must incorporate if they do not want to be left behind in a growing world concerned with environmental consciousness [7].

4.3 *Pre- and Post-pandemic Effects*

Recent crises such as COVID-19 have impacted consumer attitude toward sustainability in luxury fashion, which change the way consumers tend to approach purchasing behaviors. Studies show that there is a shift toward greater concern with the quality and longevity of a product and the ways in which it follows sustainable practices; evidence of this is the enhanced understanding of the global predicament during the COVID-19 crisis. As noted, in 2022, there has been a 31% increase of international works of luxury brands affecting China's web sales because of consumer behaviors and preferences for sustainable luxury products [22]. Post pandemic, there has been a noticeable change in consumer preferences as they have started spending money on luxury items as they are becoming more aware of the environmental impacts of their purchases [30]. Also, the pandemic intensified the trend of online shopping that lets luxury brands interact with customers who pursue sustainable activities [34].

4.4 *Segment Differences*

Segment differences imply the allocation of consumers with similar preferences and values, thereby giving rise to certain types of consumers. Among them, there a "zero-waste warriors" people for whom ecological characteristics of products are important and who strive to do as little harm to the environment as possible. According to these consumers, they look forward to buying brands that can practice efficient waste management strategies and manufacture products that are recyclable [35]. Another segment is the "transparency trackers" who pay focus on stock ethical qualities of products, including labor relations, supply chain integrity, and general credibility of the products. It is a large group that is sensitive to the ethical aspect of their purchases and believes in complete disclosure of the process of manufacturing products. By knowing these segment differences, brands are in a position to formulate specific marketing communication appeals and product segment offers that will target these specific consciousness motivations, hence maintaining and creating brand loyalty and consequent sales among these groups of shoppers [36].

Figures 3 and 4 show a detailed explanation of the segment of consumers (with respect to age) that are interested in sustainable products. Figure 4, which is the global consumer insights pulse survey conducted during the year 2021, clearly depicts how the "core millennials" and the "mature millennials" are the group that are most conscious about purchasing sustainable products among the larger crowd.

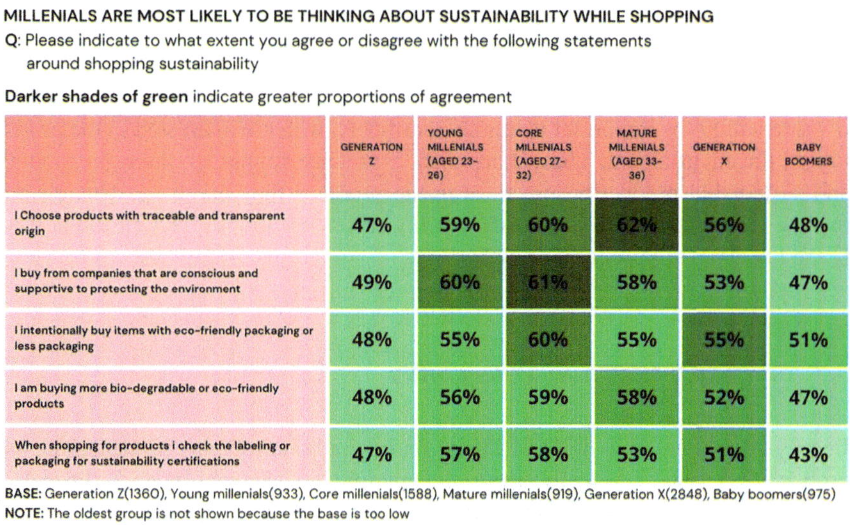

MILLENIALS ARE MOST LIKELY TO BE THINKING ABOUT SUSTAINABILITY WHILE SHOPPING

Q: Please indicate to what extent you agree or disagree with the following statements around shopping sustainability

Darker shades of green indicate greater proportions of agreement

	GENERATION Z	YOUNG MILLENIALS (AGED 23-26)	CORE MILLENIALS (AGED 27-32)	MATURE MILLENIALS (AGED 33-36)	GENERATION X	BABY BOOMERS
I Choose products with traceable and transparent origin	47%	59%	60%	62%	56%	48%
I buy from companies that are conscious and supportive to protecting the environment	49%	60%	61%	58%	53%	47%
I intentionally buy items with eco-friendly packaging or less packaging	48%	55%	60%	55%	55%	51%
I am buying more bio-degradable or eco-friendly products	48%	56%	59%	58%	52%	47%
When shopping for products i check the labeling or packaging for sustainability certifications	47%	57%	58%	53%	51%	43%

BASE: Generation Z(1360), Young millenials(933), Core millenials(1588), Mature millenials(919), Generation X(2848), Baby boomers(975)

NOTE: The oldest group is not shown because the base is too low

Fig. 3 Generational differences in eco-conscious consumerism [37]

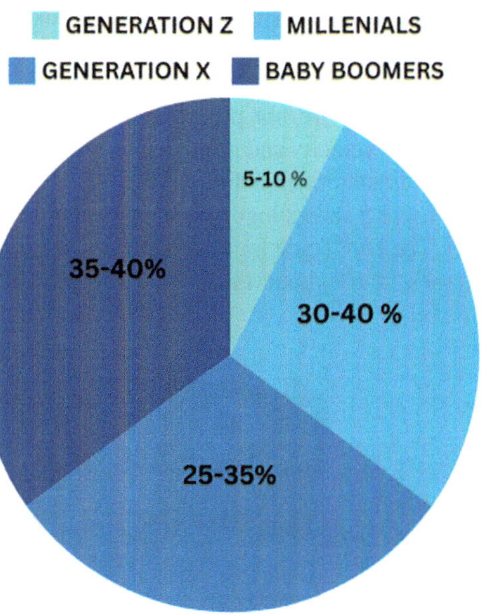

Fig. 4 Report on the different luxury consumers by age and how much each generational group consumes [38]

A recent study by McKinsey & Company found that 66% of global consumers are willing to pay more for sustainable products. This trend is particularly strong among Millennials and Gen Z consumers, who are more likely to be environmentally conscious [39].

5 Communicating Sustainability to Luxury Consumers

The analysis of a selected luxury brand has shown that the proper communication of sustainability is important to attract consumers, who are interested in the protection of the environment. These consumers appreciate simplicity, the right conduct, and sustainable production practices in the products they purchase. This chapter explores major themes of how sustainability communications affect customer trust, brand loyalty, and long-term growth.

5.1 Transparency and Traceability

Luxury consumers now demand clarity on how a brand obtains its materials, and produces its products, as well as the truth regarding its sustainability goals. Providing customers with comprehensive and easily understandable information about the consequences of product production and use contributes to achieving the recognition of brands as environmentally friendly and oriented to the formation of a tolerant society [40].

Certifications like Global Organic Textile Standard (GOTS) and OEKO ensure that a brand complies with sustainability standards, among others Fair Trade. For instance, GOTS stands for organic textile production and OEKO-TEX confirms that our products are not produced from unhealthy chemicals. Certifications such as these add veracity and believability to brand-targeted consumers who look for proof of the brand's sustainability [41].

Figure 5 shows logos of prominent certifications in sustainable and ethical fashion. The Fair Trade logo on the left represents equitable trading practices, ensuring fairness throughout the supply chain. The GOTS logo in the center signifies

Fig. 5 Logos representing key certifications in sustainable and ethical fashion [40, 41]

compliance with organic textile standards, covering ecological and social criteria. On the right, the OEKO-TEX logo certifies that textiles are free from harmful substances, promoting safety and sustainability. Together, these certifications highlight key aspects of ethical and environmentally conscious fashion.

Inclusion of blockchain technology in the supply chain augments transparency by creating a record of every process that happens in the supply chain [42]. Companies such as Everledger have used blockchain to offer auditable sources for their products and services to help consumers check whether the products were produced under the right environment and standards. This level of accountability has become one of the biggest marketing tools for brands targeting sustainability-conscious luxury consumers [26].

5.2 *Emphasizing Value Propositions*

The consumers of luxury products consequently look for values over and above the aesthetical ones, especially the sustainability of the products. Stress on the advantages for the company and the consumer, as well as on ethical aspects of long-term use of sustainable luxury products, increases their appeal. Timeless designs and longevity of products are something that consumers look for when purchasing a luxury clothing item. Products with guaranteed heavy duty, those with long-term appeal to the customer, are those that stand a better chance of appealing to those deeply embedded consumer values. Brands make sustainable clothing products by manufacturing durable clothes that ensure that people only purchase the necessary clothes, hence promoting sustainable fashion [43].

Health factor is another factor on why sustainable luxury products should be bought because the raw materials used are natural organic products. They eliminate pests, but they are unhealthy for end users who prefer natural fabrics such as cotton and linen, which do not contain chemicals and pesticides. Adopting these aspects of health benefits can attract consumers who would like to promote the sustainability of their environment as well as enjoy good health [44].

5.3 *Marketing Campaigns*

There are five main areas relevant to sustainability in luxury fashion: still, the suggested research analysis shows that some of the most important messages for sustainability can be effectively disseminated through the appropriate marketing campaigns focusing on the values and practices as well as the storytelling about luxury brands. Eco-fashion marketing is essential for effectively communicating the values of a sustainable luxury brand. By emphasizing the use of eco-friendly materials, ethical production processes, and responsible sourcing, brands can create a narrative that resonates with consumers seeking both exclusivity and environmental responsibility. Through carefully crafted storytelling and high-quality visuals, eco-fashion marketing highlights the brand's commitment to sustainability while maintaining the allure of luxury. This approach helps connect with conscious consumers who prioritize ethical choices without compromising on style or quality. By integrating sustainability into the brand's identity, eco-fashion marketing reinforces trust, builds deeper connections, and positions the brand as a leader in the evolving luxury landscape.

Influencer collaborations are important to the development of successful sustainable luxury brands because they help get the word out about brands that offer luxury goods that are also sustainably produced at a high level. That is why working with like-minded influencers will allow these brands to promote a caring approach to the environment, using nonsustainable materials and resources, and fair manufacturing. Influencers, therefore, assist in relaying these values, making the brand's message not only memorable but also more familiar [45].

This approach not only builds up the credibility of the brand but also develops a greater bond of consumers with the brand who are conscious about environmental conservation and also builds up brand loyalty and sustainability in the long run in the saturated luxury market [46].

5.4 Industry Partnerships and Collaborations

There are two critical partnerships: first, with luxury brand owners, and second, with suppliers and other members of the fashion sector involved in working to promote sustainability. They are not only beneficial to the innovation, but, more importantly, are in a way that the partners are held accountable for the sustainability practices. This is because brands unite to demand and set high standards of sustainability in sourcing procurement and overall impact on the environment.

Collaborations with material suppliers and groups interested in sustainability allow the brands to find more advanced sustainable barriers. For instance, partnerships between luxury fashion companies and textile producers creating biodegradable fabrics, or organic materials as substitutes for animal skin (Mylo™, a leather-like made from fungi), are now popular. These alliances enable brands to provide ecological solutions without eradicating luxurious and aesthetic design [47]. To increase sustainability in fashion, collaborations between different brands, nongovernmental organizations, and industry associations (for instance, Fashion for Good or the Sustainable Apparel Coalition) are essential. These are usually driven by circular economy, recycling, and low carbon economy principles. For instance, it is understood that the Kering Group works with other related sustainable organizations to invent the regenerative agriculture system for the procurement of raw materials [11].

6 Case Studies of Successful Sustainable Luxury Brands

Stella McCartney best suits luxury fashion brands that have adopted sustainability into their systems followed by Burberry and Patagonia. Stella McCartney started with environment-friendly fabric and environment-friendly policy with a vision to be a net-zero company by 2040 [48]. Burberry is also a good example, which has done a lot to overcome the problem; it has recently started by eliminating animal fur from its productions and has controlled the sources of leather, 77% of which are purchased from the tanneries that have received the corresponding certificate [49]. The clothing company Patagonia takes political stances for the environment and uses recycled materials and looks to extend the life span of products and minimize wasteful replacements [50]. Combined, these brands prove exciting and sustainable and can share harmony amid the continuity of conscientious consumerism.

6.1 *Stella McCartney*

Stella McCartney was one of the first designers to adapt a sustainable approach into her supply chain and the whole fashion business [51]. Recycling is one of the pillars of sustainability communication for the brand—and in 2022, the brand worked and produced its first circular product, a mono material parka with Econyl, 100% regenerated nylon from pre- and post-consumer waste, such as fishing nets and industrial plastics [52]. The brand is recycling all textiles that could be used; in 2012, 34.3 metric tons of waste excluding hazardous materials was diverted from landfill and recycled or reused. Additionally, every single outlet of Stella McCartney has recycling stations [53]. Currently, Stella McCartney is an excellent example of how one can combine traditional luxury with environmentally friendly production. Thanks to her, ethical luxury fashion does not conform to the norms of fashion traditionalism while also creating expectations for the next generations of designers. Given the industry's progression toward ethically efficient attaining of luxury McCartney's impact, it is expected to remain influential going forward [20].

Figure 6 demonstrates the sustainability index of the brand Stella McCartney. Their sustainability score being 7.4 is an average of the ratings of all other parameters such as transparency and environmental impact.

6.2 *Patagonia*

In terms of environmental responsibility, it is very hard to leave Patagonia out of this list. The US retailer has always specialized in outdoor and adventure wear and had a very strong focus on sustainability [55]. Early in 1996, by means of solar-tracking skylights, efficient lighting, and radiant heating, energy consumption had already

Fig. 6 Sustainability index of Stella McCartney [54]

been cut by 60% at Patagonia. It also led and explored the use of post-consumer recycled content and evaluated the company's usage of hazardous substances. It also moved from using conventional cotton to organic cotton use to 100%. It aims to hit carbon neutrality across its business by 2025, including its entire supply chain. The company is actively investing in renewable energy and has incentives that encourage employees to drive less [56]. The brand is B Corp certified and publishes annual scorecards. Patagonia supports fair trade programs for workers. The 2020 Fashion Index Report says Patagonia is one of the few brands that disclose how many workers receive a living wage.

Figure 7 represents survey respondents' association levels of characteristics "Corporate Social & Environmental Responsibility," "Fashionable & Innovative Designs," and "Products Quality & Durability" of Brand "Patagonia" in comparison to "The North Face" and "Columbia" on a scale of 0 to 5 (0 = no association, 5 = high association); 95% confidence interval bar chart. This chart clearly shows that Patagonia is doing a better job in being socially and environmentally responsible, creating fashionable and innovative designs and also good-quality products that are durable in comparison to other sustainable luxury fashion brands.

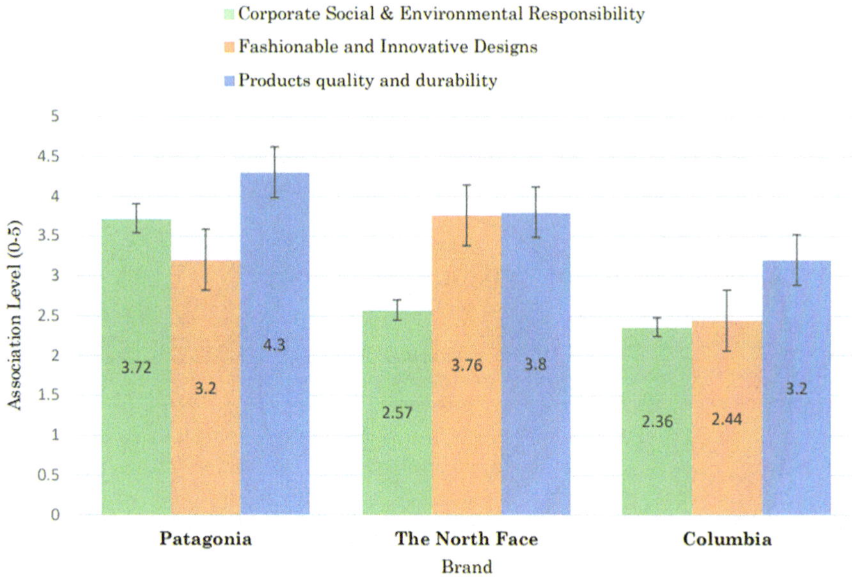

Fig. 7 Association levels of brand attributes [57]

6.3 Burberry

Burberry is a multinational brand operating in three regions: Asia Pacific (41% of retail/whole revenue), Europe, the Middle East, India, and Africa (EMEIA: 36%), and the Americas (23%) [58]. The responsibility at Burberry is reinvented from the TBL concept, which defines its mission of contributing to the improvement of the industry and the communities, as well as the evaluation of impacts from the environment. Figure 8 represents the responsibility goals regarding sustainability set by Burberry for the year 2022 [59]. The TBL is a framework that integrates three principles of organizational performance, which includes economic, environment, and social. These dimensions are also called the three Ps: people, planet, and profits [60].

The brand marked innovation at its offices, stores, warehouses, and internal manufacturing sites, effective resource management within the manufacturing process, and the use of sustainable sources for cotton, cashmere, and leather. The poly usually lays down a useful policy of responsible sourcing that enhances the conservation of natural resources, proper treatment of animals, and respect for the rights of people. Altogether with Gucci and other essential luxury brands, Burberry revealed that they will stop using animal furs in their products, which is a more sustainable textile [61].

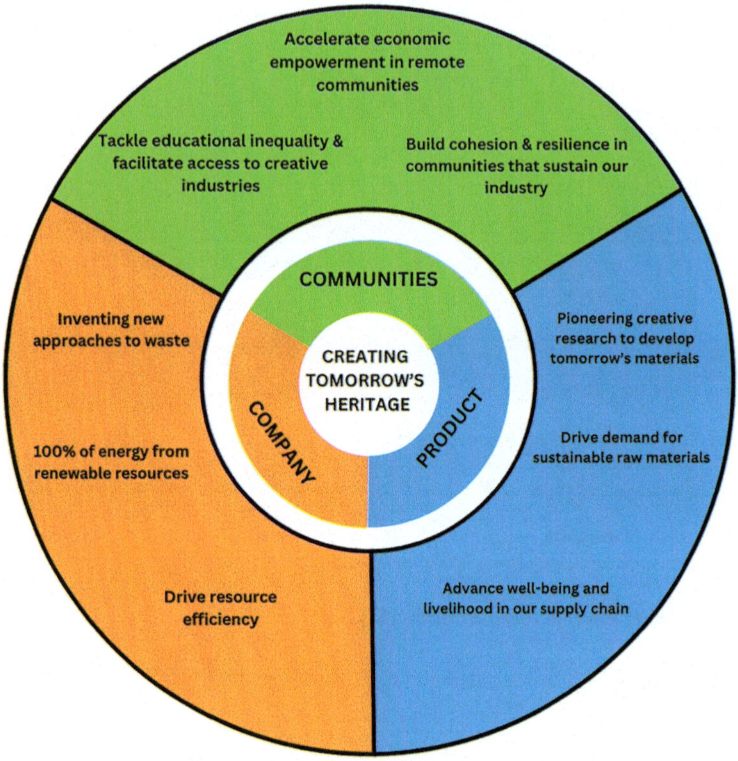

Fig. 8 Responsibility goals for 2022 from Burberry [59]

7 Conclusion

It is important to understand that the textile industry, specifically luxury clothing, is changing at a rapid pace as consumers pay attention to sustainability. It further explains that sustainability or environmentally conscious customers are not exceptional people but are the emerging market of luxury fashion. These consumers have a high expectation of luxurious products and believe that the luxurious products should embrace factors such as transparency, ethical sourcing, and use of sustainable materials. As this shift rises, brands have no option, but to align with new approaches like circular fashion design, the incorporation of responsible materials, and collaborating with partners who embrace the natural environment. This makes communication of such efforts crucial in ensuring that customers who are concerned with environmental issues remain loyal to the business. The brands that would adapt well to such a changing environment will incorporate sustainability not merely as a factor of fashion but rather as the actual foundation for business solutions to reduce the negative impact of luxury fashion.

References

1. Abdelmeguid, A., Tsironis, G., Afy-Shararah, M., Tsagarakis, K., & Salonitis, K. (2024). Insights into sustainability and circular economy trends in luxury fashion: A LinkedIn topic modelling study. *Journal of Innovation and Knowledge, 15*, 100228. https://doi.org/10.1016/j.clrc.2024.100228
2. Pittaway, D. (2020). *Sustainability in luxury fashion: Top brands and their sustainable practices*. Appnova. https://www.appnova.com/sustainability-in-luxury-fashion-top-brands-and-their-sustainable-practices/
3. Kapferer, J. N. (2016). The challenges of luxury branding. In *The Routledge companion to contemporary brand management* (pp. 473–491). Routledge. https://doi.org/10.4324/9781315796789-43
4. López, B., Rangel-Pérez, C., & Fernández, M. (2023). Sustainable strategies in the luxury business to increase efficiency in reducing carbon footprint. *Journal of Business Research, 157*, 113607. https://doi.org/10.1016/j.jbusres.2022.113607
5. Joy, A., Sherry, J. F., Venkatesh, A., Wang, J., & Chan, R. (2012). Fast fashion, sustainability, and the ethical appeal of luxury brands. *Fashion Theory, 16*(3), 273–295. https://doi.org/10.2752/175174112X13340749707123
6. Zhao, L., Peng, J., & Yu, S. (2023). Sustainable luxury and consumer purchase intention: A systematic literature review. *SAGE Open, 13*(4). https://doi.org/10.1177/21582440231216285
7. Alghanim, S., & Ndubisi, N. O. (2022). The paradox of sustainability and luxury consumption: The role of value perceptions and consumer income. *Sustainability (Switzerland), 14*(22), 14694. https://doi.org/10.3390/su142214694
8. Kim, N., & Lee, K. (2023). Environmental consciousness, purchase intention, and actual purchase behavior of eco-friendly products: The moderating impact of situational context. *International Journal of Environmental Research and Public Health, 20*(7), 5312.
9. Davies, I. A., Lee, Z., & Ahonkhai, I. (2012). Do consumers care about ethical-luxury? *Journal of Business Ethics, 106*(1), 37–51. https://doi.org/10.1007/s10551-011-1071-y
10. Wiedmann, K.-P., & Nadine Hennigs, A. S. (2009). Value-based segmentation of luxury consumption behavior. *Psychology and Marketing, 26*(7), 625–651. https://doi.org/10.1002/mar.20292
11. Achabou, M. A., & Dekhili, S. (2013). Luxury and sustainable development: Is there a match? *Journal of Business Research, 66*(10), 1896–1903. https://doi.org/10.1016/j.jbusres.2013.02.011
12. Oracle, C. (2021). *10 Reasons why Patagonia is the world's most responsible company*. Change Oracle. https://changeoracle.com/2021/09/10/10-reasons-why-patagonia-is-worlds-most/
13. Kunz, J., May, S., & Schmidt, H. J. (2020). Sustainable luxury: Current status and perspectives for future research. *Business Research, 13*(2), 541–601. https://doi.org/10.1007/s40685-020-00111-3
14. Kapferer, J. N. (2009). *The luxury strategy: Break the rules of marketing to build luxury brands*. Kogan Page Publishers. https://www.researchgate.net/publication/281251957_The_Luxury_Strategy_Break_the_Rules_of_Marketing_to_Build_Luxury_Brands
15. Goor, D. (2024). *Eco-chic: How to balance indulgence and responsibility*. London Business School. https://www.london.edu/think/can-high-end-brands-balance-indulgence-and-responsibility
16. Virginia Rolling, A. S. (2018). Are sustainable luxury goods a paradox for millennials? *Social Responsibility Journal, 14*(4), 802–815. https://doi.org/10.1108/SRJ-07-2017-0120
17. Balconi, M., Sebastiani, R., Galeone, A. B., & Angioletti, L. (2020). Sustainability in the fashion luxury branding. Using neuroscience to understand consumers' intentions towards sustainable eco-luxury items. *Neuropsychological Trends, 27*, 65–74. https://doi.org/10.7358/NEUR-2020-027-BAL1
18. Rahaman, M. T., Pranta, A. D., Repon, M. R., Ahmed, M. S., & Islam, T. (2024). Green production and consumption of textiles and apparel: Importance, fabrication, challenges and

future prospects. *Journal of Open Innovation: Technology, Market, and Complexity, 10*(2), 100280. https://doi.org/10.1016/j.joitmc.2024.100280
19. Ashkenazi, J. (2022). *Eco-fashion must-haves for spring*. JustLuxe. https://www.justluxe.com/lifestyle/leisure/feature-1973183.php
20. Bajekal, N. (2023). *Stella McCartney on wanting to make fashion more sustainable*. Time. https://time.com/6302562/stella-mccartney-sustainability-interview-lvmh/
21. Hill, M. (2024). *What is circular fashion?* Good on You. https://goodonyou.eco/what-is-circular-fashion/
22. Yang, H., Su, X., & Shion, K. (2023). Sustainable luxury purchase behavior in the post-pandemic era: A grounded theory study in China. *Frontiers in Psychology, 14*, 1260537. https://doi.org/10.3389/fpsyg.2023.1260537
23. Bhatt, T. (2023). *The influence of circular fashion in luxury fashion brands*. Blily.In. https://blily.in/the-influence-of-circular-fashion/?srsltid=AfmBOoppqvGBft1lUkWJ0FwvSK_AxJfhp00ki3uTSBFwFbE5_uuVIlio
24. Radhakrishnan, S. (2022). Circular economy in textiles and fashion. In *Sustainable approaches in textiles & fashion*. https://www.researchgate.net/publication/357700430_Circular_Economy_in_Textiles_and_Fashion
25. Adegeest, D.A. (2024). *Luxury brands are turning to blockchain for authenticity and insurance*. Fashionunited. https://fashionunited.com/news/fashion/luxury-brands-are-turning-to-blockchain-for-authenticity-and-insurance/2024021358384
26. Al-Issa, N., & Marsela Thanasi-Boçe, O. A. (2022). Boosting luxury sustainability through Blockchain technology. In *Blockchain technologies in the textile and fashion industry* (pp. 17–46). Springer. https://doi.org/10.1007/978-981-19-6569-2_2
27. Rahul Kumavat, H. C. (2023). *The impact of influencer marketing on consumer behaviour: Effectiveness on increasing brand awareness, improving brand perception, and driving sales*. https://www.researchgate.net/publication/372787352_THE_IMPACT_OF_INFLUENCER_MARKETING_ON_CONSUMER_BEHAVIOUR_EFFECTIVENESS_ON_INCREASING_BRAND_AWARENESS_IMPROVING_BRAND_PERCEPTION_AND_DRIVING_SALES
28. Wang, X., Sung, B., & Phau, I. (2024). Examining the mediating role of consumer desire for luxury: Can perceived sustainability and natural rarity evoke willingness to pay more? *Journal of International Consumer Marketing, 36*(4), 301–319. https://doi.org/10.1080/08961530.2024.2333756
29. D'Adamo, I., & Lupi, G. (2021). Sustainability and resilience after COVID-19: A circular premium in the fashion industry. *Sustainability, 13*(4), 1861. https://doi.org/10.3390/su13041861
30. Tarnanidis, T. K. (2024). Fashion consumers' response to the COVID-19 pandemic and today's future. *International Journal of Science and Research Archive, 12*(2), 648–655. https://doi.org/10.30574/ijsra.2024.12.2.1286
31. Seo, Y., & Buchanan-Oliver, M. (2015). Luxury branding: The industry, trends, and future conceptualisations. *Asia Pacific Journal of Marketing and Logistics, 27*(1), 82–98. https://doi.org/10.1108/APJML-10-2014-0148
32. Atkinson, S. D., & J. K. (2021). New luxury: Defining and evaluating emerging luxury trends through the lenses of consumption and personal values. *Journal of Product and Brand Management, 31*(3), 377–393. https://doi.org/10.1108/JPBM-09-2020-3121
33. Nguyen, K. T., Nguyen, P. N. D., Van Nguyen, L. T., Nayak, R., & Nguyen, T. Q. (2023). The practice of sustainable fashion of luxury boutique fashion brands in Vietnam: What go right, and what go wrong. *PLoS One, 18*(10), 1–18. https://doi.org/10.1371/journal.pone.0287429
34. Okur, N., Saricam, C., Iri, A. R., & Sari, I. (2023). Analyzing the impact of Covid-19 on sustainable fashion consumption with a model based on consumer value perceptions. *Journal of Fashion Marketing and Management, 27*(5), 826–850. https://doi.org/10.1108/JFMM-10-2021-0272
35. King, S. (2023). *14 zero waste clothing Brands for a Sustainable Closet*. Sustainably-Chic. https://www.sustainably-chic.com/blog/zero-waste-clothing-brands

36. Chang, W. Y., Taecharungroj, V., & Kapasuwan, S. (2022). Sustainable luxury consumers' preferences and segments: Conjoint and cluster analyses. *Sustainability (Switzerland), 14*(15). https://doi.org/10.3390/su14159551
37. PwC. (2021). The global consumer: Changed for good. In *PwC's June 2021 global consumer insights pulse survey*. https://www.pwc.com/gx/en/consumer-markets/consumer-insights-survey/2021/gcis-june-2021.pdf
38. Villalón, P. G. (2024, March 22). *Report on the different luxury consumers by age and how much each generational group consumes*. Luxonomy. https://luxonomy.net/report-on-the-different-luxury-consumers-by-age-and-how-much-each-generational-group-consumes/
39. Tudor, A. (2024). *The future of fashion: The rise of eco-conscious Brands in the Luxury Market*. The World Reporter. https://www.theworldreporter.com/2024/08/the-future-of-fashion-the-rise-of-eco-conscious-brands-in-the-luxury-market.html
40. Jestratijevic, I., Rudd, N. A., & Uanhoro, J. (2020). Transparency of sustainability disclosures among luxury and mass-market fashion brands. *Journal of Global Fashion Marketing, 11*(2), 99–116. https://doi.org/10.1080/20932685.2019.1708774
41. Staff, E. (2023). *Which sustainable certifications for clothing are Most trustworthy?* EcoCult. https://ecocult.com/eco-friendly-ethical-sustainable-labels-certifications-clothing-fashion/
42. Chirag. (2023). *How Blockchain Technology is Transforming the Fashion Industry?* Appinventiv. https://appinventiv.com/blog/blockchain-in-fashion-industry/
43. Sun, J. J., Bellezza, S., & Paharia, N. (2021). Buy less, buy luxury: Understanding and overcoming product durability neglect for sustainable consumption. *Journal of Marketing, 85*(3), 28–43. https://doi.org/10.1177/0022242921993172
44. Sebastian, A. (2024). *Eco-friendly materials for clothing: A sustainable fashion guide*. Seamapparel. https://seamapparel.com/blog/eco-friendly-materials-for-clothing/
45. Bryła, P., Chatterjee, S., & Ciabiada-Bryła, B. (2022). The impact of social media marketing on consumer engagement in sustainable consumption: A systematic literature review. *International Journal of Environmental Research and Public Health, 19*(24), 16637. https://doi.org/10.3390/ijerph192416637
46. Gartlan, D. (2024). *How green marketing can build brand loyalty*. Stevens & Tate. https://stevens-tate.com/articles/digital-content-marketing-strategies-for-manufacturer-2-2/
47. Rognoli, V., Petreca, B., Pollini, B., & Saito, C. (2022). Materials biography as a tool for designers' exploration of bio-based and bio-fabricated materials for the sustainable fashion industry. *Sustainability: Science, Practice and Policy, 18*(1), 749–772. https://doi.org/10.1080/15487733.2022.2124740
48. Pavarini, M. C. (2023, January 20). *How Stella McCartney heads for net-zero carbon in 2040*. The Brands. https://www.the-spin-off.com/news/stories/The-Brands-How-Stella-McCartney-heads-for-Net-Zero-carbon-in-2040-16930
49. Sullivan, K. (2022, November 30). *Burberry sets fire to items that animals were killed for PETA*. PETA. https://www.peta.org/blog/burberry-burns-its-unsold-items/
50. Change Oracle. (2024, January 5). *10 reasons why patagonia is the world's most responsible company*. https://changeoracle.com/2021/09/10/10-reasons-why-patagonia-is-worlds-most/
51. Lolli, A. (2023). *Stella McCartney: Fashion and sustainability*. FG Conscious Fashion. https://thefashionglobe.com/stella-mccartney
52. Birch, K. (2023). Stella McCartney: Forging the future of sustainable fashion. *Sustainability Magazine*. https://sustainabilitymag.com/sustainability/stella-mccartney-forging-the-future-of-sustainable-fashion
53. Guzman, K. (2023). *Designer Stella McCartney discusses sustainability and the future of fashion*. Yale School of Management. https://som.yale.edu/story/2023/designer-stella-mccartney-discusses-sustainability-and-future-fashion
54. Sustainability Directory Index. (2024, July 3). *Stella McCartney → Apparel & Fashion → consumer goods → Index*. Sustainability Directory. https://sustainability-directory.com/index/consumer-goods/apparel-fashion/stella-mccartney/

55. Jeremy. (2021). *Patagonia case study: How to drive profit with purpose*. Misio. https://hq.misio.io/patagonia-case-study-how-to-drive-profit-with-purpose/
56. Kammen, D. M., Hendricks, P., Pendleton-knoll, S., & Strand, R. (2018). *Patagonia's path to carbon neutrality by 2025*. The Regents of the University of California.
57. Pagnucco, G. (2023). Impact of sustainable branding on consumer decision-making: The case of Patagonia. In *Bachelor thesis: International bachelor economics and business economics*. https://thesis.eur.nl/pub/67576/THESIS-Giulia-Pagnucco-.pdf
58. Akeroyd, J. (2023). *Burberry: Annual report 2023/24*.
59. Burberry. (2023). *Burberry: Responsibility data appendix 2023/24*.
60. Bals, L., & Tate, W. (2017). *Implementing triple bottom line sustainability into global supply chains*. Routledge. https://www.routledge.com/Implementing-Triple-Bottom-Line-Sustainability-into-Global-Supply-Chains/Bals-Tate/p/book/9781783533510
61. Yotka, S. (2018). *Burberry's future is fur-free—And much more sustainable*. Vogue. https://www.vogue.com/article/burberry-fur-free

Eco-Design Principles in Luxury Clothing and Apparel

Kavitha S. and Benitta Christy P.

1 Introduction

In today's world, the importance of textiles is known to everyone for their functionality, and at the same time, everyone is aware that this industry creates a lot of problems for the environment [1]. The textile industry is the key to creating pollution although it makes a significant contribution to the Indian economy. Textile manufacture involves crucial processes that utilize a lot of chemicals through various stages of production and act as a main concern in creating air, land, and water pollution. Disposal of waste and microfiber on the environment and overconsumption of energy, chemicals, and water contribute to detrimental impacts on the environment [2]. Everyone is aware that water and chemicals are the main sources for fiber cultivation and textile wet processing [3] and its resultant wastewater containing hazardous chemicals, suspended solids, and heavy metals is directly discharged into lakes, ponds, rivers, and so on, creating incredible impact to human health, aquatic life, and environment [4–9]. The textile industry consumes high energy for production that emits close to 10% of greenhouse gas [10] and their extended supply chain increases carbon emissions [11], particularly in developing countries [12] and the wastewater generated due to dyeing and finishing is the major contributor for pollution [13]. The increase in population and fast fashion trend create demand for clothing, which results in the establishment of new textile and apparel manufacturing industries that act as base for ecological damage and climatic change [14].

Kavitha S.
Department of Home Science, Mother Teresa Women's University,
Kodaikanal, Tamil Nadu, India

Benitta Christy P. (✉)
Department of Costume Design and Fashion, PSG College of Arts & Science,
Coimbatore, Tamil Nadu, India

Sadhna et al. (eds.), *Crafting Sustainability in Luxury Textiles for a Zero-Waste Future*, Sustainable Textiles: Production, Processing, Manufacturing & Chemistry, https://doi.org/10.1007/978-3-031-86773-6_10

Today, over the past practices of textile industries in India, the consumer has started realizing the need to shift from using synthetic apparel. The consumer's behavior and approach have changed to support and use fabrics and garments made out of natural fiber and natural dye. Based on the need for natural sustainable clothing, textile manufacturers and industrialists are giving due importance to environmental sustainability issues and producing eco-friendly clothing [15], but this concern is only a micro level among the individual organization and needs to get extended at the macro level in long run [16]. On one side, societal economic development is important and could not avoid new innovations through advanced technology and industrial development but at the same ecological deterioration needs to be considered [17]. Emissions such as CO_2, nitrogen oxides, sulfur dioxide, and so on occur due to the use of fossil fuels in the textile industry [18]. Shift to nature will decrease environmental pollution as only 2.1 kg of CO_2 is emitted for producing a cotton T-shirt and 5.5 kg of CO_2 emission for the production of a polyester T-shirt [19].

Today's necessity is to overcome the pollutants deposition on air, water, and soil caused by sequential textile production processes. Currently, the textile industry urges to incorporate ethics in textile manufacturing and is transformed to produce sustainable and eco-friendly textile products that will harmonize the environment and societal development [20].

2 Early Textile and Apparel Production Practices and Environmental Repercussion

Cotton spinning prevalence is noted from historical proofs and has been a dominant fiber cultivated since 3000 BC. Being a traditional natural seed fiber, the consumption of cotton and usage increases gradually over the years. The properties of cotton are more favorable for consumers, and it is cultivated in 80 countries worldwide, holds the world market nearly up to 60% [21], and has been raising its production rate for the past 30 years [22]. Traditional cotton fiber production notably indicates that the use of pesticides, water for the purpose of irrigation, and demand for fertilizers are high and consequently the damage to society and the environment is also high.

Indian history also comments on the use of wool and silk [23] where silk is reared by killing moth cocoons and un-winded manually using a cheaper workforce [24]. Chemicals, water, and soap solutions are required to remove or clean sericin. Similar to silk, wool is a natural and protein fiber. It has its unique comfort and functional properties. Wool is obtained by removing hind skin and the tactics involved in removing the quantity of wool create lots of pain to the animals (as labor charges are made based on the quantity or amount of wool removed from animals and not on the basis of time). Wool also needs cleaning of impurities and grease using water and other required cleaning agents [25].

Synthetic or manufactured fibers are technically developed by using solutions that enter through designed tiny holes, incorporating preferred properties (size, shape, color, form, length) and which use a high amount of energy [26]. On one hand, synthetic fiber production (using dry/melt spinning) does not affect the environment as it uses a limited quantity of water and chemicals than natural fibers, but, on the other hand, over-utilizing more energy, it releases some dangerous chemicals during the manufacture of nylon, that is, ozone-depleting substance [27] and polyester carcinogenic substance during recycle [28].

Also at the end of the nineteenth century, manmade fiber (Viscose) was manufactured by making use of unwanted cotton or wood pulp. Therefore, there are chances for viscose to retain the drawbacks of cotton production. The raw material wood is very natural and involves no pesticides, but, of course, removing trees to full length will affect animal life and soil deterioration. More specifically, the wet spinning of cuprammonium rayon drives pollution as it contains some deadly chemicals [29].

2.1 Type of Manufacturing Processes

The raw material or fibers are used to convert it into the next stage (Yarn), and during spinning, the production environment is completely surrounded by microfibers in the air (causes eye irritation and breathing discomfort/lung disease) and filled with a high volume of sound/noise between 70 and 100 dB [30], which leads to hearing loss and some other fitness problem.

The subsequent stage of making yarn/thread into fabric is weaving/knitting, which creates very huge noise (beyond the permitted limit) because of the use of older, outdated, poorly designed machinery [31]. Also, dust in the workplace (very small fiber particles float in the air) may be inhaled and impact human health. The manufactured fabrics undergo wet processing and finishing, of which the chemicals used during sizing (pentachlorophenol), desizing, scoring (fiber contaminants into water as waste), bleaching (whitening using sodium hypochlorite raise water waste with BOD), Mercerization (discharges caustic soda into wastewater) dyeing/printing (raise of BOD/COD lead to carcinogenic amines), and textile finishes (releases deadly chemicals) are discharged into water as waste and, once when it ends up into the ocean, create harmful effects to human health and ecosystem [19].

The transformation of the clothing industry is quick, and with the advent of fast fashion, economic growth, technological inventions, family income level, the rise in population, and purchase level, the industry is one of the major contributors to environmental pollution. Worldwide, entire production networks of the clothing industry of USD 1.3 trillion offer employment to nearly 300 million human forces. The consumer of the apparel fashion industry practices overconsumption of clothing in line with following latest trends [32] which leads to higher carbon emissions. The apparel industry poses serious environmental issues due to consumer purchase behavior as it grows day by day. It was stated that the consumer's apparel purchase during 2014 was nearly 60% higher than their apparel purchase made in the year

2000 [33]. The fashion seasons are from 2 cycles (traditional world) to 50 cycles (Fast fashion World). If the consumption pattern continues in the same direction, it may increase over three times in 2050.

3 Understanding the Facts and Data About the Hitches of Textile and Apparel Industry

For the selection of any resources, one needs to understand the consumption level of energy, chemical utilization, and concern for humans and the environment. Cotton is the dirtiest pesticide-ridden crop [34], creates water toxicity [35–37], and uses more kinds of fertilizers and dangerous chemicals (nitrogen, phosphorus, potassium, insecticides, etc.) of which nitrogen (harmful to the ecosystem) discharges N_2O that is more dangerous than greenhouse gases [38]. Insecticides used for cotton are very dangerous as it is responsible for creating severe health risks of damaging the nervous system and reproductive syndrome and to the core can even kill a person when not properly exposed to the crop. Cotton cultivation practices using pesticides and fertilizers are subjected to emit CO_2 and SO_2 which exceeds the synthetic polyester production [39].

The production of wool is organic without chemical inclusion, but goats are fed vegetation which has been reported to spoil air quality from the Pacific to North America [40]. Silk is known for its royal appeal and has pride all over the world. However, the key issue with silk filament extraction is the destruction of pupa (during stifling) when exposed to hot air/steam. This was protested by several animal welfare organizations. The other considerable issues with silk are using of more energy (for transportation, stifling, and cacoon boiling), water pollution (uses chemicals for silk cleaning), and the requirement for huge land resources to grow mulberry trees [29].

The commonly used synthetic fibers after cotton usage are polyester and nylon [41], but production (thermal processing) of those fibers causes severe infections to eyes, nose, and throat [42] and emits nitrous oxide (i.e., 0.2% in air/atmosphere) in which one-tenth of 0.2% is because of the production of these two synthetic fibers at a large. Nitrous oxide has the ability to damage the ozone layer (stratospheric) 300 times more than carbon dioxide.

The ginning process (separating the cotton fiber from seeds) engenders huge noise (practice of using high-speed machinery) given in Tables 1 and 2 and air pollution [45]. The workers were exploited to work for long hours under high noise (caused by doffing brush with 500–1000 cycles/s) than the standard noise frequency level of 31.5–250 cycles/s [46], causing temporary/permanent hearing loss to workforce [47, 48]. A study reveals that, in Maharashtra, the employees worked in high noise levels of 89–106 dBA [49, 50]. Also, generation of fine and micro-level fiber/cotton dust particles in the ginning room (approximately 2000–6000 $\mu g/m^3$ density of dust particles) causes severe respiratory diseases and occupational disorders in the workforces [3, 51–54].

Table 1 Noise exposure time and limit for industry work persons [43]

Exposure time (in hours)	Limit in < dB(A)
8	90
4	93
2	96
1	99
½	102
1/8	108
1/16	111
1/32 (2 min) or less	114

Table 2 Noise exposure permissible limit of OSHA [44]

Duration per day (h)	OSHA 1910.95 (dBA)
8	90
6	92
4	95
3	97
2	100
1½	102
1	105
½	110
<1/4	115

Similar to the ginning process even during spinning (yarn formation process), a variety of high-speed machinery is employed that creates huge noise and generates miniature/tiny dust particles in the workplace. At least 30% of the workforce is getting exposed to both dust/dirt (i.e., 14 times high) and noise (i.e., 90 dBA) that exceeds the industrial standard limits [55–58], and due to noise and air pollution, the workforces are imposed to serious health issues such as hearing impairment, respiratory problems, nervous disorders, occupational lung disorders, and psychic disorders [59–61].

Weaving is a fabric formation process and is one of the noisiest (Tables 3 and 4) producing segments by the use of pitiable, very old-fashioned looms that create terrible noise (shuttle loom) during loom shed, that is, between 85 and 104 dBA [63] and weaving industries usually exceed their allowed limit not only in India but also in other countries, that is, Karachi, Thailand, Tanzania, and Kenya with 88.4–104 dBA, 101.3 ± 2.7 dB, 92–103.8 dBA, and 99–101 dBA, respectively [64–67].

Fabric wet processing is measured to be one of the critical stages that implicate a lot of chemicals, synthetic dyes, alkalis, softeners, acids, and so on to undergo the fabric preparatory process, that is, sizing and desizing, scouring (to pull out dirt particles), bleaching and mercerization (to advance color, water penetrability), coloration (to achieve good design/patterns by dyeing/printing), and finishing processes (enhancement of functionality, look, feel and touch). Almost all raw materials applied on fabric are synthetic and chemical substances that are non-biodegradable and more poisonous. Of these chemicals, nearly 20% of unutilized/unfixed dyes and auxiliaries

Table 3 Allowable level of noise for textile industries [62]

Section	Noise level, dBA
Ginning	88–92
Blowroom	80–83
Carding	84–89
Draw frames	84–88
Speed frames	82–86
Ring frame	86–90
Rotor	85–100
Winding	82–86
Warping	80–86
Sizing	73–86
Loom shed (no-auto)	94–99
Loom shed (auto)	95–97
Shuttle loom	99–104

Table 4 Cotton textile processing and standard limit along the actual level [63]

	Exposure limit		
Defined body	Processing stage	Standard limit (μg/m^{3})	Actual level (μg/m^{3})
OSHA PEL TWA	Yarn manufacturing	200	1900–2700
	Weaving	750	1820–1960
	Ginning and waste recycling	1000	2000–6000

come out as waste/effluent [68–70] and stand as a chief contributor to pollution and create a risk to human and aquatic life [71–74]. This process poses severe damage to the stratospheric ozone layer (due to the release of carbon tetra chloride and carbon dioxide) and is venomous (due to the escalation of BOD and COD) for animals living in water [75–77]. Chemical processed effluent containing water on evaporation pollutes the air. It was noted that the discharged chemical waste into the Bandi River, Rajasthan, and the Kaduna River, Nigeria, polluted the air [78–80].

Among many industry/factory maneuvers, the apparel industry is a hard, labor-intensive [81], and valued industry as global exports of garments in a year cross over US$310 billion [82]. It functions diversely in different sites within or outside the country to execute work orders. This industry was holding a huge Indian market from 2019 to 2024 (CAGR~3–4%), except for the year 2021 due to the pandemic. A lot of factors were responsible for the progress of the apparel or clothing industry's domestic retail market. During FY2024, its export market was around USD 34.4 billion as domestic market exports also had a setback during 2021 due to the COVID-19 pandemic.

It is a well-known fact that the apparel industry next to the oil industry stands back as a great polluter of utilizing more (approximately 93 billion cubic meters) water (Table 5) and microfiber (approximately 500,000 tons) and is accountable for (between 8% and 10%) global carbon release [84–86]. Under the working environment, the apparel industry exploits the labor force by non-maintenance of workers' safety, excess work exploitation, and low and minimum wage for work [87, 88].

Table 5 Fabric wet process and its water consumption level with effluent level [83]

	Water consumption		Effluent
Activities	Variation	Average	Variation
(a) Sizing/slashing	50–820	435	50–80
(b) Desizing	250–2100	1175	250–2100
Bleaching			
(c) Yarn (hypochlorite)	2400–4800	3600	2250–4600
(d) Yarn (H_2O_2)	2400–3200	2800	2250–3050
(e) Cloth (hypochlorite)	4000–4800	4400	3800–4600
(f) Cloth (H_2O_2)	1700–3200	2450	1700–3200
(g) Mercerizing	3600–17,000	10,600	3500–17,500
Dyeing			
(h) Yarn (light & medium shade)	3600–4800	4200	3500–4700
(i) Yarn (dark shade)	4800–6400	5600	4700–6300
(j) Yarn (very dark shade)	6600–8800	7700	6500–8700
(k) Cloth (light and medium shade)	7800–9600	8700	7700–9500
(l) Cloth (dark shade)	10,400–12,800	11,600	10,300–12,700
(m) Cloth (very dark shade)	14,300–17,600	15,950	14,200–17,500

Table 6 Fiber type and unfixed dyes [90]

Fiber type	Dye type	Unfixed dyes
Wool and nylon	Acid dyes, reactive dyes for wool	7–20%
	Premetalised dyes	2–7
	After chromes	1–2
Cotton and viscose	Azoic dyes	5–10
	Reactive dyes	20–50
	Direct dyes	5–20
	Pigment	1
	Vat dyes	5–20
	Sulphur dyes	30–40
Polyester	Disperse	8–20
Acrylic	Modified basic	2–3
Polypropylene	Spun dyed	N/A

Records prove that the textile and apparel industry is growing huge year by year. As the market grew very big, the buying purchase behavior among consumers is also increasing, and their social intention to buy more clothing proves their least botheration for environmental concerns [89]. The consciousness about environmentally friendly apparel is less and thus consumer increases their buying behaviors [90, 91]. It was noted that the consumption of clothing in 2014 was higher than the consumption level in 2000. Textile and clothing production is closely associated with water usage and has a direct impact on polluting water (Table 6) and thus needs to control human existence. Among many industries, the textile and apparel industry alone pays a huge for pollution and it is time to take a diversion to shift toward new and innovative business models that concern for environment and human safety.

4 Waste Generation and Causes

Management of waste in the textile industry costs very high. Textile waste arises out of post-industrial operations and after consumer usage. Textile production waste arises in various stages such as during spinning (dust, leaves, seeds, lint fibers), weaving (leftover yarns in warp beam and cones), knitting (fault during loop formation, yarn stains, yarn variations), dyeing (shade difference), printing (mismatch of repeat), processing, and finishing (chemical and solid waste).

Today, the lifestyle has changed, and based on every function/occasion, the attires are wearied. It is one such factor that, in a single day, two to three sets of garments are used by an individual, which motivates the manufacturers to open more textile industries and increase their production rate. The people's attitude toward using textiles transformed from Indian Culture (using a piece of fabric to its maximum lifetime/generation to generation) to the influenced Western culture (less usage and disposal of fabric before its lifetime). Fast fashion is one such factor that increases global warming as it upraises consumers to purchase more garments than they require. Also, the frequency of clothing purchases has increased largely. This kind of consumer attitude boosts the production operation of textile and clothing industries at a higher scale.

Textile wet processing involves increased usage of water and chemicals in all pre-treatments and finishing (dyeing/printing). The chemical-dyed fabrics hold toxins on disposal, increase landfill, and spoil the environment [92, 93]. The textile industry follows the concept of utilizing fiber/raw materials, making fabrics, using the product, and disposal of textiles after usage. In a very short span, consumers move to new products to adopt the latest trends. As a whole, the approach may be called, in short, Take-Make-Use-Dispose, and when this system gets repeated, it causes severe pollution, raises landfill, and ends up in environmental degradation [94, 95].

Environmental pollution grows with technological development. This is due to a lack of awareness about eco-concepts and a lack of sustainable practices among the population. The consumer should be made aware that second cloth, reusing, or forwarding the textiles to others will stop falling of waste textiles into landfill. Lack of knowledge of product design and development results in a quick landfill. During product designing, all necessary information is to be sourced, reviewed, and planned appropriately so as to retain the product for a longer duration and to keep it suitable for all fashion trends.

5 Strategies for Extending Sustainability in the Textile Industry

It is a crucial and serious moment to consider and refine all strategies and technologies used in textiles industries that are sustainable and environment friendly.

5.1 *Environment-Friendly Product Design*

To produce an environment-friendly product, the primary essential step is planning to source the resources that are eco-friendly. Also employing of process that utilize less energy is necessary, structured for the type of product to be manufactured to meet consumer preference. The distribution channel/mode of product reaches consumer, usage functions, disposal/reuse behavior, and reduction of environmental impact throughout its lifecycle is also concentrated. This principle activity helps to integrate and correlate the production process with the low impact on ecology [96].

5.2 *Ecological Practices in Spinning*

The spinning manufacturer should think about the possibilities of developing/producing an ecological product/yarn that will be highly appreciable rather than thinking of recycling the product to reduce pollution and energy consumption. High-end latest energy-saving machinery and use of renewable energy sources could be used for spinning yarn which naturally helps to reduce carbon footprint. For example, regular waste cleaning of suction tubes, replacement of energy-efficient motors used in carding, replacement of old air conditioners with energy-efficient new air conditioners (with proper duct plates, increase in air cycles and circulation), and care for compressor maintenance all contribute to reduce and save energy utilization.

The waste that arises during spinning also uses energy, and thus the spinning waste could be reused in the blow room again (either utilized directly or reprocessed waste for next usage). The use of biodegradable/reusable packing cones, packing materials, and storage materials such as bobbins and cans will encourage sustainable practice [12] and decrease carbon emissions/pollution. Earlier polyethylene bags have been used for packing yarns which could be replaced by Hydrocarbon-free jute sacking [97–99].

5.3 *Ecological Practices in Weaving*

In earlier practice, during weaving, a lot of resources are depleted, it utilizes more chemicals/energy, and it generates a high level of waste.

Based on Fig. 1, it is a known fact that a lot of energy is used during warping, sizing, and weaving. This could be replaced by the use of energy-efficient machinery for warping and weaving (net-zero emissions). Also, the use of recyclable/biodegradable materials for sizing will motivate ecological progression. Correct maintenance of humidity in workplaces will limit the density of

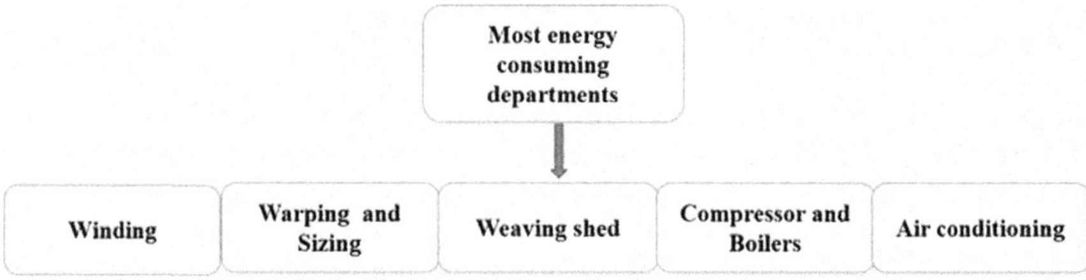

Fig. 1 Weaving and their energy-consuming areas

floating cotton dust and optimum circulation of air per hour aids/promotes workers' safety.

5.4 *Ecological Practices in Fabric Pre-treatments and Finishing*

Pre-treatment of fabric is essential to improve fabric characteristics and thus unavoidable but needs to be undertaken carefully by the use of enzyme technologies for processing of fabrics (enzymatic desizing, enzymatic scouring, and enzymatic bleaching), use of natural dyes and mordants, green and harmless chemicals, less use of energy, and wastewater management (zero liquid discharge (ZLD)). In comparison to conventional methods, the use of ozone for processing (bleaching) limits the use of water [100] and one-step processing (desizing–scouring–bleaching) may comparatively reduce the effluent [101]. ZLD aims not to leave any liquid as waste and to utilize treated water to the fullest to improve water usage efficiency.

Dyeing utilizes more chemicals and auxiliaries, but modifying the dye structure/type of dye/dyeing technology/equipment may reduce ecological impacts. For example, if the fiber structure is changed, then a better scope of dyeability will be attained (without salt usage) through exhaust dyeing [102]. Spray dyeing and low liquor ratio dyeing lessen water usage [103]. Currently, polyester fabrics dyed with supercritical fluid are very economical and limit environmental pollution. Alternatively, the ultrasonic process removes inside air and improves dye molecule penetration and diffusion.

The conventional printing process is not ecological and uses unsustainable and toxic materials/chemicals. As it implies defined designs and extraordinary opacity inks the print resembles to be durable and definite. Automated machines help to reduce energy level conception during the printing process. Avoidance of plastic inks and use of water-based printing ink (certainly no chemical additives) may reduce harmfulness. Gentle drying by hanging the fabric is preferred for tumble dryers. Recently, direct-to-garment print, otherwise called digital printing, is getting popular as it fuses very fast with fibers after print (saves energy, is fast, accurate, and realistic design print, is cheap, is adaptable to all kinds of fabrics, and prints more colors).

As with textile pre-treatments, the finishing of textile materials/fabrics also involves more and more chemicals to modify fabric characteristics and appearance. Plasma treatment on textile materials is an ecological method used for changing both natural and synthetic fabric surface properties/characteristics [104–106]. This technology is used on wool to restrict shrinkage and on polyester to increase super-hydrophobicity and other properties [107–109]. Application of enzymes (i.e., natural extract of renewable resources) in moderate condition usage, which is a non-polluting chemical, easy to manage and control the pollution is an ecological practice but costlier when compared to textile chemical pretreatments [110].

5.5 *Ecological Practices in Garment Manufacturing*

The adoption of sustainable goals [111] and principles [112] in apparel manufacturing will enable and lead the industry to move on the right path. All approaches start from suppliers relation, designing of products (eco-design), ecological production (zero waste) [113], waste reduction (recycling/reuse) [114], and resource and environment management [115–118].

Today, due to consumer awareness/want for ecological clothing, clothing manufacturers are invited to produce garments/apparel implementing sustainable approaches. The producers source raw materials that are recyclable and organic (cotton, bamboo, hemp, recycled polyester, lyocell, etc.)

Proper selection of resources (men, machine, and material) should be utilized appropriately by having proper control and measures to reduce wastage, that is, sourcing of the right supplier, inventory maintenance, staff training, durable eco garment design, and efficiency during marker making, cutting, and sewing. Energy management (energy efficient machinery and the use of alternate/renewable energy sources) during apparel production is very important to cut down costs and conserve resources. For smooth and prolonged industry operation, fair labor and business/trade practices should be adopted [119]. Garment waste disposal raises the problem of environmental landfill, and to reduce the same, proper waste management and disposal is to be practiced. Even during the packing and transport of apparel, the industry should ensure that all materials and activities are ecological and follow sustainable practices throughout their supply chain.

6 Circular Business Approaches

The earlier textile industry used linear business models for business development with the least botheration of utilizing resources and energy at maximum level, which resulted in high emissions and waste discharge. Instead, the circular business model/approach is more concerned about very less use of energy and resources, showing greater interest in developing quality and life-extending/

Table 7 Ecological movements for circular economy with digital enablers [125, 126]

Operation	Digital enabler	Effect on circular economy
Repair	3D printing	Extended product life, conserving resources, and minimizing waste
Reverse logistics	Traceability devices	Recovered material for reuse, repair, or recycling aiding value creation in the textile industry
Product as a service (PaaS)	Traceability devices	Efficient use of resources and discouraged product ownership resulting in a low carbon footprint
Recycling and sorting	Near-infrared technology	Accurate sorting of waste for material recovery and value creation. Increased material flow and decreased use of virgin materials

durable products, extreme use of single garment for varied functionalities, and extension of garment usage by more than one person through resale/reuse, limiting the environment landfill through recycle [120]. To employ a circular economy, businesses should adopt digitalization to arrive at correct and prompt data for making decisions to improve sustainability [121], reduce the gaps in the process [122], show new views for maximum resource utilization [123], increase transparency and traceability, and motivate supply chain management [124]. The application of artificial intelligence helps in regulating the production process as given in Table 7 [127].

7 Life Cycle Assessment

Life cycle assessment (LCA) acts as a major approach/tool that facilitates and assists textile companies in reducing the environmental impact—recognizing the scope, understanding of feed material, assessment of end product, and clarification to know the reduction of environmental impact. LCA helps industry persons analyze each and every stage of a process or the entire cycle of a product that travels through, that is, from fiber sourcing to disposal/reuse [128]. This analysis supports the industrialist to assess their existing operations and move to the next stage of making decisions (in case of environmental impact) to proceed further [129] toward their sustainable practices [130–134]. Very famous concerns (Nike and Levi Strauss & Co) led LCA to understand their products and their impact on the environment [135]. The LCA of LEVI'S 501® jeans (Table 8) proved that recycling jeans saves energy, reduces resource consumption, and limits pollution rather than producing new jeans.

Methodologies of LCA: The carbon footprint of products is measured from the entry of the company gate to the exit of the company gate, and it has no botheration for how the raw materials were produced/extracted and moved till their gate [136].

Table 8 Findings of life cycle impact analysis for LEVI'S 501® jeans [136]

		Fiber	Fabric assembly	Cut, sew, finish	Sundries & packaging	Transport logistics retail	Consumer care	End of life	Total
Climate change (kg CO_2-e)	Value	2.9	9.0	2.6	1.7	3.8	12.5	0.9	33.4
	Share	9%	27%	8%	5%	11%	37%	3%	100%
Water consumption (liters)	Value	2565	236	34	77	10	860	0	3781
	Share	68%	6%	1%	2%	0%	23%	0%	100%
Eutrophication (PO_4-e g)	Value	18.0	5.5	2.9	7.9	3.1	7.9	3.5	48.9
	Share	37%	11%	6%	16%	6%	16%	7%	100%
Land occupation (m^2/year)	Value	9.3	0.2	0.0	0.5	0.3	1.7	0.0	12.0
	Share	78%	1%	0%	4%	2%	14%	0%	100%
Abiotic depletion (mg Sb-e)	Value	19.9	7.2	1.9	118.5	4.4	17.9	0.1	29.1
	Share	12%	4%	1%	70%	3%	11%	0%	100%

8 Discussion and Conclusion

Today, various sustainable methods are practiced in the textile industry with the agenda of filtering the waste that falls into land as landfill; rather, we circulate the materials on the need to address very seriously that all alternative methods/technologies adopted are completely biodegradable. For example, currently digital printing is substituting conventional printing and is called sustainable practice, but still, the ink used is synthetic which needs to be replaced by natural printing ingredients.

The textile industry is a great contributor to pollution and currently as an industry requirement, and textile manufacturers through sustainable practices (eco-friendly product design and manufacturing, energy-efficient technologies, following waste reduction and management strategies, and ethical practices) satisfy consumer demands. One such notable ecological development is ZLD as water is the source for many of the textile processes. The industry is transforming green due to a lot of research and sustainable practices. Consumer awareness toward adopting green fabrics and apparel has grown gradually which urges manufacturers to produce eco-friendly apparel products and their attitude toward disposing of used textiles and apparel is also improved. Today, consumers concerning ecological health adopt post-consumer textile disposal practices (recycle, reuse, and refashion).

Eco-designing of textile products is possible and will be successful only by associating the knowledge of manufacturers, designers, policymakers, researchers, and ultimately consumers. The apparel and fashion business extends in two lines of track (i.e., fast fashion and slow fashion). Comparatively slow fashion is considered to be gentle and steady which streamlines material selection, production process, level of apparel collections, and finally high reputation and lengthier value for their garment designs. Slow fashion can otherwise be called ethical fashion. The youth young people are more affectionate toward the latest trends and follow fast fashion [137]. The fashion industry and media could divert this fashion trend among youngsters by focusing on ethical/eco-friendly fashion [138].

Changing over to circular design strategies and principles from the linear business model will drive forward aiming to increase product durability and recyclability. However, this kind of shift involves many challenges that could be solved by joint efforts of industry players, policymakers, stakeholders, Government, and consumers. Continuous research efforts, innovative technologies, social responsibilities, and consumer awareness about environmental pollution are very vital for the development of the textile and apparel industry. Educating consumers about sustainable fashion and the practice of sustainable technologies is the key to overcoming environmental challenges. Ethical business practices, LCA, and digital transformation will accelerate the industry.

References

1. Rathore, B. (2022). Textile Industry 4.0 transformation for sustainable development: Prediction in manufacturing & proposed hybrid sustainable practices. *Eduzone: International Peer Reviewed/Refereed Multidisciplinary Journal, 11*(1), 223–241.
2. Niinimäki, K., Peters, G., Dahlbo, H., Perry, P., Rissanen, T., & Gwilt, A. (2020). The environmental price of fast fashion. *Nature Reviews Earth & Environment, 1*(4), 189–200.
3. Gulzar, T., Farooq, T., Kiran, S., Ahmad, I., & Hameed, A. (2019). Green chemistry in the wet processing of textiles. In *The impact and prospects of green chemistry for textile technology* (pp. 1–20). Woodhead Publishing.
4. Abro, Z. A., Chen, N., Yifan, Z., Cheng-Yu, H., Abassi, A. M. R., Simair, A. A., et al. (2018). Investigation on thermal comfort characteristics of regenerated bamboo and cotton woven structured fabrics. *Autex Research Journal, 18*(4), 323–329.
5. Allan Heap, S., Greenwood, P. F., Leah, R. D., Eaton, J. T., Stevens, J. C., & Keher, P. (1983). Prediction of finished weight and shrinkage of cotton knits—The starfish project: Part I: Introduction and general overview. *Textile Research Journal, 53*(2), 109–119.
6. Gita, S., Hussan, A., & Choudhury, T. G. (2017). Impact of textile dyes waste on aquatic environments and its treatment. *Environment and Ecology, 35*(3C), 2349–2353.
7. Berradi, M., Hsissou, R., Khudhair, M., Assouag, M., Cherkaoui, O., El Bachiri, A., & El Harfi, A. (2019). Textile finishing dyes and their impact on aquatic environs. *Heliyon, 5*(11), e02711.
8. Abdullahi, L. I., Yakasai, M. A., Bello, B., & Khan, A. R. (2016). Determination of heavy metals and acute toxicity studies of vat dyes on earthworm (Lumbricusterrestris) as ecological risk indicators. *IOSR Journal of Pharmacy, 6*(2), 31–36.
9. Angelis-Dimakis, A., Alexandratou, A., & Balzarini, A. (2016). Value chain upgrading in a textile dyeing industry. *Journal of Cleaner Production, 138*, 237–247.
10. UNFCCC. (2018). *UN helps fashion industry shift to low carbon*. United Nations Framework Convention on Climate Change.
11. European Parliament. (2021). *The impact of textile production and waste on the environment (infographic)*. European Parliament.
12. Chen, X., Memon, H. A., Wang, Y., Marriam, I., & Tebyetekerwa, M. (2021). Circular economy and sustainability of the clothing and textile industry. *Materials Circular Economy, 3*, 1–9.

13. Morlet, A., Opsomer, R., Herrmann, S., Balmond, L., Gillet, C., & Fuchs, L. (2017). *A new textiles economy: Redesigning fashion's future*. Ellen MacArthur Foundation.
14. Berg, A., Granskog, A., Lee, L., & Magnus, K.-H. (2021). *Fashion on climate—How the fashion industry can urgently act to reduce its greenhouse gas emissions*. McKinsey and Company and Global Fashion Agenda.
15. Bharagava, R. N., & Chowdhary, P. (Eds.). (2019). *Emerging and eco-friendly approaches for waste management* (pp. 245–269). Springer.
16. Azizi, M., Mohebbi, N., & De Felice, F. (2016). Evaluation of sustainable development of wooden furniture industry using multi criteria decision making method. *Agriculture and Agricultural Science Procedia, 8*, 387–394.
17. Figge, F., Givry, P., Canning, L., Franklin-Johnson, E., & Thorpe, A. (2017). Eco-efficiency of virgin resources: A measure at the interface between micro and macro levels. *Ecological Economics, 138*, 12–21.
18. Ruan, X., Ding, N., & Yang, J. (2022). Dual-level eco-efficiency analysis in sustainable management: An empirical study on textile manufacturers and supply chains in China. *Journal of Environmental Management, 304*, 114227.
19. Hasanbeigi, A., & Price, L. (2015). A technical review of emerging technologies for energy and water efficiency and pollution reduction in the textile industry. *Journal of Cleaner Production, 95*, 30–44.
20. Kirchain, R., Olivetti, E., Miller, T. R., & Greene, S. (2015). *Sustainable apparel materials*. Materials Systems Laboratory, Massachusetts Institute of Technology.
21. de Oliveira, C. R. S., da Silva Júnior, A. H., Mulinari, J., & Immich, A. P. S. (2021). Textile re-engineering: Eco-responsible solutions for a more sustainable industry. *Sustainable Production and Consumption, 28*, 1232–1248.
22. Kadolph, S. J., & Langford, A. L. Textiles 2002
23. Textiles Intelligence. (2006). Textiles Intelligence
24. Mitchell, V. (2004). Mary Schoeser, World textiles: A concise history. *Textile, 2*(1), 108–111.
25. Khatri, A., Peerzada, M. H., Mohsin, M., & White, M. (2015). A review on developments in dyeing cotton fabrics with reactive dyes for reducing effluent pollution. *Journal of Cleaner Production, 87*, 50–57.
26. Talukdar, M. K. (2001). Noise pollution and its control in textile industry. *Indian Journal of Fibre & Textile Research, 26*(1), 44–49.
27. Bhaduri, G., & Ha-Brookshire, J. (2017). The role of brand schemas, information transparency, and source of message on apparel brands' social responsibility communication. *Journal of Marketing Communications, 23*(3), 293–310.
28. Gupta, S., Gwozdz, W., & Gentry, J. (2019). The role of style versus fashion orientation on sustainable apparel consumption. *Journal of Macromarketing, 39*(2), 188–207.
29. Yates, D. (1994). Organic cotton. *Green Alternatives, 4*(1), 33–36.
30. Sharma, M. (2021). Indian history of textile and apparel sector. *International Journal for Modern Trends in Science and Technology, 7*, 360–367.
31. Leal Filho, W., Perry, P., Heim, H., Dinis, M. A. P., Moda, H., Ebhuoma, E., & Paço, A. (2022). An overview of the contribution of the textiles sector to climate change. *Frontiers in Environmental Science, 10*, 973102.
32. Islam, S. U., & Butola, B. S. (2018). *The impact and prospects of green chemistry for textile technology*. Woodhead Publishing.
33. Laperrire, L., & Reinhart, G. (2014). *CIRP Encyclopedia of production engineering*. Springer Publishing Company.
34. Sathiyaraj, S., Kumar, A. S., & Subramani, A. K. (2015). Consumer perception towards online grocery stores, Chennai. *Zenith International Journal of Multidisciplinary Research, 5*(6), 24–34.
35. Chen, H. L., & Burns, L. D. (2006). Environmental analysis of textile products. *Clothing and Textiles Research Journal, 24*(3), 248–261.
36. McDonough, W., & Braungart, M. (2002). *Transforming the textile industry*.

37. Kumari, P., Saroj, S., Singh, J., & Neelam, M. R. (2013). Eco—textiles: For sustainable development. *International Journal of Scientific & Engineering Research, 4*(4), 2229–5518.
38. Rana, S., Pichandi, S., Parveen, S., & Fangueiro, R. (2014). Natural plant fibers: Production, processing, properties and their sustainability parameters. In *Roadmap to sustainable textiles and clothing: Eco-friendly raw materials, technologies, and processing methods* (pp. 1–35). Springer.
39. Bhar, C. (2016). Indian textile industry and its impact on the environment and health: A review. *International Journal of Information Systems in the Service Sector, 8*(4), 33–46.
40. Karthik, T., & Rathinamoorthy, R. (2017). Sustainable silk production. In *Sustainable fibers and textiles* (pp. 135–170). Woodhead Publishing.
41. Laird, W., & Baker, R. V. (1982). Confirmation of sources of cotton gin noise by multivariate analysis techniques. *Transactions of the ASAE, 25*(4), 1050–1056.
42. Kumar, R. S. (2014). *Process management in spinning*. CRC Press.
43. Roy Choudhury, A. K. (2014). Environmental impacts of the textile industry and its assessment through life cycle assessment. In *Roadmap to sustainable textiles and clothing: Environmental and social aspects of textiles and clothing supply chain* (pp. 1–39). Springer.
44. Kanwal, A., Nazeer, M. A., & Rasul, S. (2023). Digitalization in the textile sector for circularity. In *Circularity in textiles* (pp. 213–231). Springer Nature.
45. Kramer, S. B., Reganold, J. P., Glover, J. D., Bohannan, B. J., & Mooney, H. A. (2006). Reduced nitrate leaching and enhanced denitrifier activity and efficiency in organically fertilized soils. *Proceedings of the National Academy of Sciences, 103*(12), 4522–4527.
46. Dube, K. J., Ingale, L. T., & Ingale, S. T. (2011). Hearing impairment among workers exposed to excessive levels of noise in ginning industries. *Noise and Health, 13*(54), 348–355.
47. Khatik, V., Shinde, A., & Thakare, K. (2013). The Pioneer study on identification of fire hazards in cotton ginning Industries of Nandurbar Region of Maharashtra. *International Journal of Scientific Research, 2*(10), 1–4.
48. Dube, K. J., Ingale, L. T., & Ingle, S. T. (2013). Respiratory impairment in cotton-ginning workers exposed to cotton dust. *International Journal of Occupational Safety and Ergonomics, 19*(4), 551–560.
49. Woldeyohannes, M., Bergevin, Y., Mgeni, A. Y., & Theriault, G. (1991). Respiratory problems among cotton textile mill workers in Ethiopia. *Occupational and Environmental Medicine, 48*(2), 110–115.
50. Jannet, J. V., & Jeyanthi, G. P. (2006). Pulmonary health status of ginning factory women laborers in Tirupur, India. *Indian Journal Of Occupational And Environmental Medicine, 10*(3), 116–120.
51. Bünger, J., Schappler-Scheele, B., Hilgers, R., & Hallier, E. (2007). A 5-year follow-up study on respiratory disorders and lung function in workers exposed to organic dust from composting plants. *International Archives of Occupational and Environmental Health, 80*, 306–312.
52. Pavan, M., Samant, L., Mahajan, S., & Kaur, M. (2024). Role of chemicals in textile processing and its alternatives. In *Climate action through eco-friendly textiles* (pp. 55–72). Springer Nature.
53. Kan, C. W. (2015). Plasma treatments for sustainable textile processing. In *Sustainable apparel* (pp. 49–118). Woodhead Publishing.
54. Amutha, K. (2017). Sustainable chemical management and zero discharges. In *Sustainable fibers and textiles* (pp. 347–366). Woodhead Publishing.
55. Planet, H. (2019). *Global environment outlook GEO-6: Healthy Planet, healthy people*. Cambridge University Press.
56. Policy, I. (2010). *A guide for developing countries*. World Bank.
57. Lee, J. (2014). *The latest in sustainable textiles*. Triple Pundit (People, Plane, Profit).
58. Ballikar, V. (2013). *Textile industry and environmental issues*. Fibre2Fashion.
59. Osnos, S. (2007). Past, present and future of continuous basalt fiber. *JEC Composites, 35*, 24–25.
60. Wishman, M., Hagler, G. E., Lewin, M., & Pearce, E. M. (1998). *Handbook of fiber chemistry*. Taylor and Francis.

61. Malloy, T., & Grubb, M. (2008). *MSDS–Nylon 6*.
62. Talukdar, M. K. (2001). Noise pollution and its control in textile Industry. *Indian Journal of Fibre and Textile Research, 26*, 44–49.
63. Hasanbeigi, A. (2013). *Emerging energy-efficiency and carbon dioxide emissions-reduction technologies for the iron and steel industry*. International Energy Analysis Department, Energy Analysis and Environmental Impacts Division.
64. Allwood, J. M., Laursen, S. E., De Rodríguez, C. M., & Bocken, N. M. (2006). *Well dressed? The present and future sustainability of clothing and textiles in the United Kingdom* (Vol. 1). University of Cambridge.
65. Bedi, R. (2006). Evaluation of occupational environment in two textile plants in northern India with specific reference to noise. *Industrial Health, 44*(1), 112–116.
66. Bhatt, S. R., Subrahmanyam, K., & Swami, K. R. (1990). *Noise pollution in textile industry: A review*. ATIRA.
67. Yhdego, M. (1991). Urban environmental degradation in Tanzania. *Environment and Urbanization, 3*(1), 147–152.
68. Picard, M., Girard, S. A., Simard, M., Larocque, R., Leroux, T., & Turcotte, F. (2008). Association of work-related accidents with noise exposure in the workplace and noise-induced hearing loss based on the experience of some 240,000 person-years of observation. *Accident Analysis & Prevention, 40*(5), 1644–1652.
69. Giardino, D. A., & Durkt, G. (1996). Evaluation of muff-type hearing protectors as used in a working environment. *American Industrial Hygiene Association Journal, 57*(3), 264–271.
70. Van Kempen, E. E., Kruize, H., Boshuizen, H. C., Ameling, C. B., Staatsen, B. A., & De Hollander, A. E. (2002). The association between noise exposure and blood pressure and ischemic heart disease: A meta-analysis. *Environmental Health Perspectives, 110*(3), 307–317.
71. Öhrström, E., Björkman, M., & Rylander, R. (1979). Subjective evaluation of work environment with special reference to noise. *Journal of Sound and Vibration, 65*(2), 241–249.
72. Ashraf, H. D., Younus, M. A., Kumar, P., Siddiqui, M. T., Ali, S. S., & Siddiqui, M. I. (2009). Students' corner-frequency of hearing loss among textile industry workers of weaving unit in Karachi, Pakistan. *The Journal of the Pakistan Medical Association, 59*(8), 575.
73. Gitau, J. M., Mwikali, J. M., Bett, K. W., & Njau, Z. G. (2004, June). *Noise induced hearing loss among textile industry workers in Eldoret, Kenya*.
74. Chavalitsakulchai, P., Kawakami, T., Kongmuang, U., Vivatjestsadawut, P., & Leongsrisook, W. (1989). Noise exposure and permanent hearing loss of textile workers in Thailand. *Industrial Health, 27*(4), 165–173.
75. Jayaraynum, J. (1991). Primary health care approach in occupational health and safely. *African Newsletter on Occupational Health and Safety, 1*, 72–75.
76. Kalliala, E., & Talvenmaa, P. (2000). Environmental profile of textile wet processing in Finland. *Journal of Cleaner Production, 8*(2), 143–154.
77. Ernest Orlando Lawrence Berkeley National Laboratory. (2013). *Potential for reducing paper mill energy use and carbon dioxide emissions through plantwide energy audits: A case study in China*.
78. Babu, B. R., Parande, A. K., Raghu, S., & Kumar, T. P. (2007). Cotton textile processing: Waste generation and effluent treatment. *Journal of Cotton Science, 11*, 141–153.
79. Das, S. (2000). *Some issues of ecological hazards in textile industry*. Global Brands Li & Fung.
80. Patel, S., Rajor, A., Jain, B. P., & Patel, P. (2013). Performance evaluation of effluent treatment plant of textile wet processing industry: A case study of Narol Textile Cluster, Ahmedabad, Gujarat. *International Journal of Engineering Science and Innovative Technology, 2*(4), 290–296.
81. Imtiazuddin, S. M., Mumtaz, M., & Mallick, K. A. (2012). Pollutants of wastewater characteristics in textile industries. *Journal of Basic & Applied Sciences, 8*(2), 554–556.
82. Jauregui, E., & Luyando, E. (1999). Global radiation attenuation by air pollution and its effects on the thermal climate in Mexico City. *International Journal of Climatology: A Journal of the Royal Meteorological Society, 19*(6), 683–694.

83. Garrett, B., Shorofsky, B., & Radcliffe, R. (2016). *Evaluation of textile treatment and treatment alternatives for the village of Jasol in Rajasthan*. Northern University.
84. Rathore, J. (2012). Studies on pollution load induced by dyeing and printing units in river Bandi at Pali, Rajasthan, India. *International Journal of Environmental Sciences, 3*(1), 735–742.
85. Lin, S. H., Kincade, D. H., & Warfield, C. (1994). Productivity and production in the apparel industry. *International Journal of Clothing Science and Technology, 6*(1), 20–27.
86. Atkar, A., Pabba, M., Sekhar, S. C., & Sridhar, S. (2021). Current limitations and challenges in the global textile sector. In *Fundamentals of natural Fibers and textiles* (pp. 741–764). Woodhead Publishing.
87. Khitous, F., Urbinati, A., & Verleye, K. (2022). Product-service systems: A customer engagement perspective in the fashion industry. *Journal of Cleaner Production, 336*, 130394.
88. Peters, G., Li, M., & Lenzen, M. (2021). The need to decelerate fast fashion in a hot climate-a global sustainability perspective on the garment industry. *Journal of Cleaner Production, 295*, 126390.
89. Manickam, P., & Duraisamy, G. (2019). 3Rs and circular economy. In *Circular economy in textiles and apparel* (pp. 77–93). Woodhead Publishing.
90. Köhler, A. R. (2013). Challenges for eco-design of emerging technologies: The case of electronic textiles. *Materials & Design, 51*, 51–60.
91. Mukherjee, S. (2015). Environmental and social impact of fashion: Towards an eco-friendly, ethical fashion. *International Journal of Interdisciplinary and Multidisciplinary Studies, 2*(3), 22–35.
92. Zheng, Y., & Chi, T. (2015). Factors influencing purchase intention towards environmentally friendly apparel: An empirical study of US consumers. *International Journal of Fashion Design, Technology and Education, 8*(2), 68–77.
93. Earley, R., & Goldsworthy, K. (2015). *Designing for fast and slow circular fashion systems: Exploring strategies for multiple and extended product cycles*. Nottingham Trent University.
94. da Silva, P. C., de Oliveira Neto, G. C., Correia, J. M. F., & Tucci, H. N. P. (2021). Evaluation of economic, environmental and operational performance of the adoption of cleaner production: Survey in large textile industries. *Journal of Cleaner Production, 278*, 123855.
95. Okafor, C. C., Madu, C. N., Ajaero, C. C., Ibekwe, J. C., Nzekwe, C. A., Okafor, C., et al. (2021). Sustainable management of textile and clothing. *Clean Technologies and Recycling, 1*, 70–87.
96. Schumacher, K. A., & Forster, A. L. (2022). Textiles in a circular economy: An assessment of the current landscape, challenges, and opportunities in the United States. *Frontiers in Sustainability, 3*, 1038323.
97. Diabat, A., Kannan, D., & Mathiyazhagan, K. (2014). Analysis of enablers for implementation of sustainable supply chain management–a textile case. *Journal of Cleaner Production, 83*, 391–403.
98. Bahtiyari, M. İ., & Benli, H. (2020). Comparison of ozone-based cold bleaching processes with conventional pretreatment of cotton. *Ozone: Science & Engineering, 42*(5), 450–460.
99. Imran, M. A., Hussain, T., Memon, M. H., & Rehman, M. M. A. (2015). Sustainable and economical one-step desizing, scouring and bleaching method for industrial scale pretreatment of woven fabrics. *Journal of Cleaner Production, 108*, 494–502.
100. Niu, T., Wang, X., Wu, C., Sun, D., Zhang, X., Chen, Z., & Fang, L. (2020). Chemical modification of cotton fabrics by a bifunctional cationic polymer for salt-free reactive dyeing. *ACS Omega, 5*(25), 15409–15416.
101. Lin, L., Zhu, W., Zhang, C., Hossain, M. Y., Oli, Z. B. S., Pervez, M. N., et al. (2021). Combination of wet fixation and drying treatments to improve dye fixation onto spray-dyed cotton fabric. *Scientific Reports, 11*(1), 15403.
102. Zille, A., Oliveira, F. R., & Souto, A. P. (2015). Plasma treatment in textile industry. *Plasma Processes and Polymers, 12*(2), 98–131.
103. Wen, L., Jeong, D. C., Javid, A., Kim, S., Nam, J. D., Song, C., & Han, J. G. (2015). Conductive polythiophene-like thin film synthesized using controlled plasma processes. *Thin Solid Films, 587*, 66–70.

104. Jeong, D. C., Javid, A., Wen, L., Choi, E. J., Park, S. Y., Kim, Y. H., et al. (2016). Low-temperature plasma polymerization of dicyclopentadiene for anti-corrosion properties. *Polymer, 92*, 133–139.
105. Fatkullina, R. R., Yarullin, R. M., Abutalipova, L. N., & Khadysov, B. A. (2022). Effect of low-temperature plasma treatment on the properties of a complex wool-containing material. *Journal of Physics: Conference Series, 2379*(1), 012002.
106. Morshed, M. N., Behary, N., Bouazizi, N., Guan, J., Chen, G., & Nierstrasz, V. (2019). Surface modification of polyester fabric using plasma-dendrimer for robust immobilization of glucose oxidase enzyme. *Scientific Reports, 9*(1), 15730.
107. Sun, Y., Ouyang, B., Rawat, R. S., & Chen, Z. (2020). Rapid and stable plasma transformation of polyester fabrics for highly efficient oil–water separation. *Global Challenges, 4*(7), 1900095.
108. Ibrahim, N. A., Amin, H. A., Abdel-Aziz, M. S., & Eid, B. M. (2022). A green approach for modification and functionalization of wool fabric using bio-and nano-technologies. *Clean Technologies and Environmental Policy, 24*(10), 3287–3302.
109. Mangano, G., & Zenezini, G. (2019). The value proposition of innovative last-mile delivery services from the perspective of local retailers. *IFAC-PapersOnLine, 52*(13), 2590–2595.
110. Arana, C., Franco, I. B., Joshi, A., & Sedhai, J. (2020). SDG 15 life on land: A review of sustainable fashion design processes: Upcycling waste organic yarns. In *Actioning the global goals for local impact: Towards sustainability science, policy, education and practice* (pp. 247–264). Springer.
111. D'Adamo, I., Lupi, G., Morone, P., & Settembre-Blundo, D. (2022). Towards the circular economy in the fashion industry: The second-hand market as a best practice of sustainable responsibility for businesses and consumers. *Environmental Science and Pollution Research, 29*(31), 46620–46633.
112. Eppinger, E. (2022). Recycling technologies for enabling sustainability transitions of the fashion industry: Status quo and avenues for increasing post-consumer waste recycling. *Sustainability: Science, Practice and Policy, 18*(1), 114–128.
113. Li, W., & Li, C. (2021). Strategic analysis of green marketing in Chinese textile and garment industry under the background of big data. In *In 2021 international conference on information technology and biomedical engineering (ICITBE)* (pp. 197–201). IEEE.
114. Koep, L., Morris, J., Dembski, N., & Guenther, E. (2021). Buying practices in the textile and fashion industry: Past, present and future. In *Sustainable textile and fashion value chains: Drivers, concepts, theories and solutions* (pp. 55–73). Springer.
115. Khandaker, S., Bashar, M. M., Islam, A., Hossain, M. T., Teo, S. H., & Awual, M. R. (2022). Sustainable energy generation from textile biowaste and its challenges: A comprehensive review. *Renewable and Sustainable Energy Reviews, 157*, 112051.
116. Nguyen, G. N., Mani, V., Kha, M. K., & Papadopoulos, T. (2023). Supply chain social responsibility in labour-intensive industries: A practitioner's perspective. *Production Planning & Control, 34*(4), 371–390.
117. https://www.white2labelmanufacturing.com/blog/sustainable-clothing-manufacturing
118. Coscieme, L., Manshoven, S., Gillabel, J., Grossi, F., & Mortensen, L. F. (2022). A framework of circular business models for fashion and textiles: The role of business-model, technical, and social innovation. *Sustainability: Science, Practice and Policy, 18*(1), 451–462.
119. Luoma, P., Penttinen, E., Tapio, P., & Toppinen, A. (2022). Future images of data in circular economy for textiles. *Technological Forecasting and Social Change, 182*, 121859.
120. Wiegand, T., & Wynn, M. (2023). Sustainability, the circular economy and digitalisation in the German textile and clothing industry. *Sustainability, 15*(11), 9111.
121. Hedberg, A., & Šipka, S. (2021). Toward a circular economy: The role of digitalization. *One Earth, 4*(6), 783–785.
122. Casciani, D., Chkanikova, O., & Pal, R. (2022). Exploring the nature of digital transformation in the fashion industry: Opportunities for supply chains, business models, and sustainability-oriented innovations. *Sustainability: Science, Practice and Policy, 18*(1), 773–795.

123. Chauhan, C., Parida, V., & Dhir, A. (2022). Linking circular economy and digitalisation technologies: A systematic literature review of past achievements and future promises. *Technological Forecasting and Social Change, 177*, 121508.
124. Rebitzer, G., Ekvall, T., Frischknecht, R., Hunkeler, D., Norris, G., Rydberg, T., et al. (2004). Life cycle assessment: Part 1: Framework, goal and scope definition, inventory analysis, and applications. *Environment International, 30*(5), 701–720.
125. Wenzel, H. (1998). Application dependency of LCA methodology: Key variables and their mode of influencing the method. *The International Journal of Life Cycle Assessment, 3*, 281–288.
126. https://link.springer.com/chapter/10.1007/978-3-031-49479-6_8
127. Henry, B. K., Russell, S. J., Ledgard, S. F., Gollnow, S., Wiedemann, S. G., Nebel, B., et al. (2015). LCA of wool textiles and clothing. In *Handbook of life cycle assessment (LCA) of textiles and clothing* (pp. 217–254). Woodhead Publishing.
128. García Sánchez, M., Padilla-Rivera, A., & Güereca, L. P. (2023). Social life cycle assessment of Mexico City's water cycle. *Advanced Sustainable Systems, 7*(6), 2300024.
129. Buckner, C. A., Lafrenie, R. M., Dénommée, J. A., Caswell, J. M., & Want, D. A. (2018). Complementary and alternative medicine use in patients before and after a cancer diagnosis. *Current Oncology, 25*(4), e275.
130. Wu, M., Sadhukhan, J., Murphy, R., Bharadwaj, U., & Cui, X. (2023). A novel life cycle assessment and life cycle costing framework for carbon fiber-reinforced composite materials in the aviation industry. *The International Journal of Life Cycle Assessment, 28*(5), 566–589.
131. Gheewala, S. H. (2023). Life cycle assessment for sustainability assessment of biofuels and bioproducts. *Biofuel Research Journal, 10*(1), 1810–1815.
132. Bianco, I., Gerboni, R., Picerno, G., & Blengini, G. A. (2022). Life cycle assessment (LCA) of MWool® recycled wool fibers. *Resources, 11*(5), 41.
133. Levi Strauss. (2015). *The life cycle of a jean: Understanding the environmental impact of a pair of Levi's 501 jeans*.
134. Li, M., & Subramaniam, B. (2017). LCA for green chemical synthesis—terephthalic acid. In *Reference module in earth systems and environmental sciences* (pp. 387–396). Elsevier.
135. Sandak, A., Sandak, J., Brzezicki, M., Kutnar, A., Sandak, A., Sandak, J., et al. (2019). Biomaterials for building skins. In *Bio-based building skin* (pp. 27–64). Springer.
136. Hammar, T., Peñaloza, D., Hanning, A. C., Haatanen, N., & Pakkasmaa, J. (2023). Life cycle assessment of textile fiber-to-fiber recycling by cellulose carbamate technology. *Journal of Cleaner Production, 426*, 139189.
137. Haug, A., & Busch, J. (2016). Towards an ethical fashion framework. *Fashion Theory, 20*(3), 317–339.
138. Brewer, M. K. (2019). Slow fashion in a fast fashion world: Promoting sustainability and responsibility. *Laws, 8*(4), 24.

Sustainable Polymers, Fibres, and Textile Chemicals in Luxury Textile Manufacturing

Pavan M, Sauda Samreen, N. G. Yashaswini, and E. Gladwin Joseph

1 Introduction

Luxury textiles are premium fabrics crafted from rare and costly natural fibres or cutting-edge materials. These high-end fabrics are favoured by leading designers to produce sophisticated wardrobe staples. What sets luxury textiles apart is their exceptional quality, exclusivity, and meticulous craftsmanship, all of which contribute to their elevated price [1, 2].

The term "luxury" originally referred to fabrics crafted from rare fibres such as silk, cashmere, or Egyptian cotton [1]. However, in modern times, the concept of luxury in textiles has shifted to emphasize design intricacy over mere composition. Luxury textiles are now distinguished by Rarity, Quality, Innovation, and Aesthetics (Fig. 1).

- **Rarity**
  These fabrics are often crafted from scarce natural fibres, making them less readily available. Certain luxury textiles are crafted from rare or limited resources, like vicuña wool or unique silk varieties. Additionally, some are produced in limited runs or boast exclusive designs, enhancing their appeal and making them highly sought after by collectors [2–4].

Pavan M (✉) · N. G. Yashaswini · E. Gladwin Joseph
Department of Fashion and Apparel Design, NITTE School of Fashion Technology and Interior Design, Bengaluru, Karnataka, India
e-mail: yashaswini.ng@nitte.edu.in; Gladwin.j@nitte.edu.in

S. Samreen
Department of Interior Design and Decoration, NITTE School of Fashion Technology and Interior Design, Bengaluru, Karnataka, India

Sadhna et al. (eds.), *Crafting Sustainability in Luxury Textiles for a Zero-Waste Future*, Sustainable Textiles: Production, Processing, Manufacturing & Chemistry, https://doi.org/10.1007/978-3-031-86773-6_11

Fig. 1 Distinguishing luxury textiles [1–7]

- **Quality**
  They represent the highpoint of craftsmanship, with exceptional attention to detail. Luxury textiles are created with an exceptional focus on detail, often incorporating intricate techniques such as hand-weaving, embroidery, or embellishments. These textiles use the highest quality fibres, which guarantee both outstanding durability and a rich, luxurious texture [1, 5, 6].
- **Innovation**
  Luxury textiles may incorporate the latest and most advanced materials. Some luxury textiles incorporate cutting-edge technologies or sustainable practices, leading to fabrics with distinct features such as moisture-wicking, breathability, or eco-friendliness. These textiles blend luxury with practicality, perfectly suited for contemporary lifestyles [2, 7].
- **Aesthetics**
  They are characterized by distinctive designs, intricate patterns, and lavish finishes. Luxury textiles are renowned for their extraordinary softness, smooth texture, and opulent feel, coupled with an elegant appearance. Their ability to drape beautifully and gracefully elevates the aesthetic appeal of any garment or product they are used in [1, 4, 6].
  The luxury textile industry encompasses fabrics adorned with embellishments such as embroidered lace, fringe, sequins, laser-cut silk or cotton, and appliqued wool. These exquisite textiles are employed in the creation of high-fashion ensembles, evening gowns, bespoke suits, upholstered furniture, and exclusive accessories [4, 6].
  Luxury textiles are no longer limited to clothing but are also widely used in accessories, home furnishings, interior decoration and even automotive interiors, where their visual and tactile qualities elevate the user experience. For example,

luxury brands use these textiles to create custom garments that not only represent status but also showcase artistry and innovative design. In high-end fashion, luxury textiles enable designers to explore new creative possibilities, resulting in distinctive pieces that often become emblematic of the brands [8].

The luxury fashion industry has recognized the growing significance of sustainability in recent years. As consumers grow more cognizant of environmental issues, luxury brands are being driven to implement sustainable practices across their supply chains. This includes responsibly sourcing materials, adopting eco-friendly production techniques, and upholding ethical labour standards. For instance, many luxury brands are now incorporating recycled materials into their collections or investing in innovative, environmentally friendly materials like organic cotton or bio-based polymers. This shift towards sustainability not only addresses consumer concerns but also strengthens brand loyalty and enhances marketability, as consumers are more inclined to support brands that reflect their values. Consequently, the role of luxury textiles is evolving, with a growing focus on sustainability that aligns with the changing expectations of consumers and the industry's commitment to environmental responsibility [9, 10].

Sustainability has become an essential factor in the production of luxury textiles as the fashion industry is subject to heightened scrutiny regarding its environmental impact. Consumers are now looking for luxury brands that prioritize ethical and eco-friendly practices [9]. Luxury fashion houses are responding by adopting sustainable initiatives, such as the following [11–13]:

- **Using organic and recycled materials**: Brands are exploring alternatives to traditional luxury fibres, such as organic cotton, recycled polyester, and innovative materials derived from algae or mushrooms.
- **Implementing circular business models**: Some luxury brands are embracing the circular economy by offering repair services, resale platforms, and take-back programmes.
- **Ensuring transparency and traceability**: Luxury labels are working to provide comprehensive information regarding the origin and manufacturing processes of their textiles, allowing consumers to make informed choices.

By prioritizing sustainability, luxury fashion brands can reduce their environmental impact while maintaining their commitment to quality and craftsmanship. According to Ellis [10], globally, the luxury textile market was valued at around $120 billion in 2023. Despite facing economic challenges and the effects of the COVID-19 pandemic, the market has maintained a compound annual growth rate of 5% for the past 5 years. Projections indicate that, by 2028, the market could reach 160 billion dollars, fuelled by increasing demand in the Asia-Pacific region and the ongoing digitalization of commerce.

2 Need for Sustainability in Luxury Textiles

Luxury textile brands encounter a complex challenge in harmonizing the enhancement of their brand prestige with a strong commitment to sustainability. Their selective and environmentally conscious business gradually expects not only high-quality products but also sustainable and ethical production practices. This scrutiny extends beyond the luxury sector, reflecting a broader trend where consumers and stakeholders alike demand that products are both safe and environmentally responsible.

The primary sustainability concerns for luxury textile brands centre on four key areas as shown in Fig. 2 [9]. For that, brands must navigate these challenges to meet the evolving expectations of their clientele while maintaining their market position and standing.

Sustainability in luxury textiles can be adopted in the following aspects of the luxury textile industry:

- **Principles of green chemistry**: It enables sustainable practices in the manufacturing of chemicals while protecting the environment. It lays great emphasis on process design that minimizes the usage of hazardous substances and hence sustains ecology, guaranteeing economic viability [14]. The amalgamation of green chemistry into luxury textiles emphasizes eco-friendly materials, processes, and finishes, aligning with consumer demand for safer and more sustainable products.

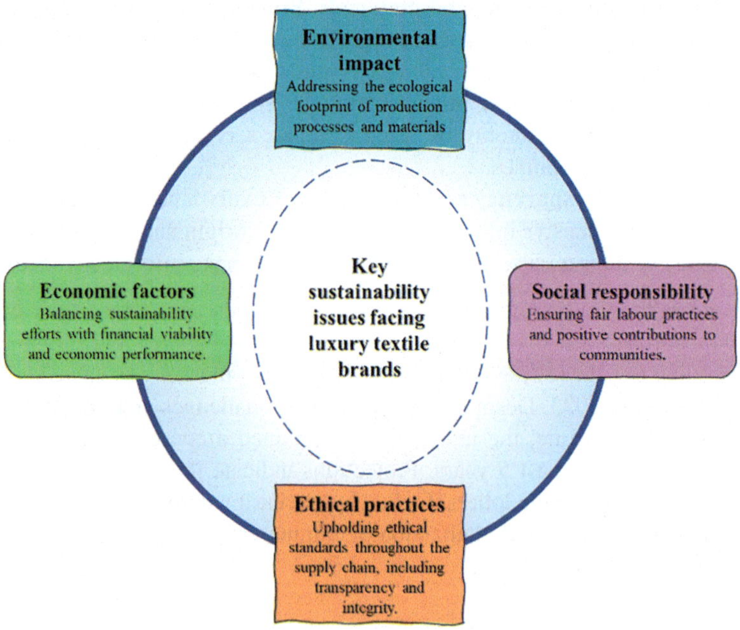

Fig. 2 Four primary sustainability concerns for luxury textile brands [9]

- **Innovations in eco-friendly dyeing and finishing processes**: Recently, several innovative processes and techniques have emerged, developed aimed at reducing water and energy consumption, as well as minimizing harmful emissions and waste disposal in the textile industries [15].
- **Bio-based and biodegradable polymers**: Shifting the focus from conventional raw materials drawn from finite resources (like polyester) and produced through harmful practices (like cotton) towards bio-based and biodegradable polymers are at high gain. This shift can revolutionize and induct sustainability in luxury textile brands [16].
- **Advanced and sustainable chemicals**: Innovating and employing advanced techniques for producing sustainable chemicals and eco-friendly treatment processes will help in substituting the use of toxic chemicals in the fabrication of luxury textile products [17].

3 Eco-Friendly Polymers and Fibres in High-End Fashion

The adoption of eco-friendly fibres and polymers is crucial in reducing the environmental impact of the luxury textile industry. As shown in Table 1, The use of natural fibres such as bamboo and organic cotton has increased in substitute of conventionally cultivated cotton, which demands a lot of water and dangerous chemicals. Particularly organic cotton is regarded as one among the top sustainable fibre because it uses less energy and water, and does not involve chemical fertilizers, herbicides, pesticides, and other synthetic chemicals during its cultivation [24, 29]. In response to the increase in consumer demand for eco-friendly products, luxury brands are taking forward innovative ways to offer the integration of sustainable fibres. This will benefit developing nations economically as well as ecologically. Furthermore, eco-friendly polymers are becoming more well-known due to their beneficial effects on the environment, which supports the trend of sustainable luxury fashion. The luxury apparel business is addressing growing worries about its environmental impact by putting more emphasis on sustainable materials, such as biodegradable alternatives [18, 29]. Under this heading, various sustainable natural, regenerated, and synthetic fibres employed in the production of luxury textiles are described.

3.1 Organic Cotton

Organic cotton is cultivated without any use of synthetic chemicals, including pesticides, regulators for plant growth, synthetic fertilizers, or defoliants. It adheres to an established organic production plan [30]. Hence, it offers environmental sustainability for healthy living and increasingly finds use in high-status luxury fabrics that possess great softness, biodegradability, and hence good uses

Table 1 Various application of sustainable fibres and polymers in luxury textiles

Natural fibres	Applications
Organic cotton [18]	Cosmeto-textiles are also used in high-added-value textiles incorporating skin and health care
	Comfort and health: It is ideal for a number of exercise, meditation, and leisure products, because it naturally keeps you in comfort and health
	Value additions: Ecological chemical and natural extract treatments provide value addition towards functionality on antimicrobial, UV protection, and self-cleaning for high-value textiles
Linen [19]	*Bedding*: Linen is especially suitable for bedding since it can absorb high moisture and is best for cool and healthy sleep-in hot climates
	Health benefits: Linen fabric contains antibacterial and antifungal properties, thus promising a cleaner sleeping environment
	Versatility: It offers both fashion and home décor due to its unique texture and appearance, catering to style and functional benefits
Hemp [21]	*Luxury fashion*: Hemp is highly popular in the sphere of luxury fashion because it reflects the company's concern for sustainability and eco-friendliness
	Interiors: This is used for high-quality, long-lasting textiles that offer a unique look in luxury interiors
	Versatility: Hemp unlocks a wide range of exciting possibilities for sustainable innovations in fashion and interior design
Pinatex leather [22]	*Fashion products*: Employed in belts, wallets, shoes, and handbags as a leather substitute, but in a more environmentally friendly manner
	Clothing: Used in the manufacturing of multi-use and environmentally friendly clothes
	Furniture: The material is good to go for green upholstery and interior decoration
	Products for pets: The product is utilized to manufacture pet leashes, hence extending the field of application
Mushroom leather [23]	*Sustainable material*: It is made from fungi; hence, it is a bio-based leather alternative and free of plastic
	Animal welfare: Reduces demand for animal leather and, consequently, the number of animal deaths attributed to the industry
	Composting: Production waste products can be composted. This will help in a circular economy
	Fashion integration: Practiced by Ashoka Paris, among many other fashion brands, it plays a great role in sustainable fashion
Bamboo fibre [24]	*Textile applications*: Bamboo fibre is popular for use in underwear, T-shirts and socks due to its breathability and moisture-absorbing qualities
	Home Textiles: The field of home textiles finds use in bathrobes, towels, bed clothes, and mats for comfort and aesthetic appeal
	Specialized clothing: Include bamboo fibres, which have antibacterial properties, and are particularly used for pregnant women and children's summer clothes to keep them away from UV radiation
	Durability: Bamboo fabrics retain anti-bacterial properties even after a number of washes, which would make bamboo a durable option to use in an apparel product

(continued)

Table 1 (continued)

Natural fibres	Applications
Milk fibre [25]	Best suited for garments that prioritize comfort and sustainability, such as underwear, loungewear, and shirts
	Milk fibre dyes easily and retains vibrant colours, making it an excellent choice for unique, high-end designs
Viscose rayon [26]	Viscose rayon finds its use in a lot of man's, women's, and children's wear manufactured by textile industries. Home textiles also use this material in curtains and bed linen. It is also useful in carpeting and in medical textiles
Lyocell [27]	Bioactive properties can be conferred on garments and technical textiles by the use of lyocell fibres. Treatment of Lyocell fibres with copper ions develops exceptional antimicrobial properties, hence extending their use in hygiene and safety applications such as antibacterial, antiviral, and antifungal textiles
Recycled plastic fibres [28]	*Apparel*: It is used in moisture-wicking apparel and hard-wearing outdoor wear
	Accessories: Include those used to make eco-friendly bags, hats, and shoes
	Home furnishings: Utilized in sustainable upholstery and carpeting
	Industrial use: Used to generate industrial materials that are strong and durable

within apparel and home textiles. With an increase in awareness of health, hygiene, and ecological issues, demand has risen significantly, extending its use to high-fashion items and other eco-friendly products. It is also often treated with environmentally friendly chemicals, making it more appealing in the higher-end textile market [18].

3.2 Linen

Linen is a textile made from the fibres of the flax plant (*Linum usitatissimum* L). In addition to its unique aesthetic and hand, linen is valued for its own merits for its high tensile properties, good absorbency, and cooling effect. Linen fabrics help in managing human body temperature, which is essential for comfort as well as well-being, while the high moisture desorption characteristic makes it dry quickly. Especially for those who reside in warm areas, linen promotes restful sleep. Although with these advantages, Linen textiles are stiff and have low elasticity; hence, it is normally blended with other fibres to make the fabric even more comfortable [19, 20].

3.3 Hemp

Hemp (*Cannabis sativa*), one of the oldest and most eco-friendly natural fibres used in fabric construction, has been utilized since around 800 BC. Hemp has less than 0.2% of tetrahydrocannabinol, which makes it non-narcotic and useful for a variety of purposes [31]. The fibres are extracted from the bast of the plant and fibres have high strength and decay resistance. The length of hemp fibres depends on the technical length of the stems, measured from the root crown to the mid-inflorescence. The most valuable primary fibres are found in the middle and upper sections of the hemp stems, while the lower parts contain fibres of lesser functionality [21, 20].

3.4 Bamboo Fibres

Bamboo fibres are those materials made from the lignocellulosic part of bamboo plant [32]. Fabric constructed from bamboo fibres is known for its silky texture, making it a desirable choice for luxury textiles [33]. Bamboo viscose is largely produced and is relatively inexpensive, but, at the same time, it has environmental downsides [32]. Bamboo contains a distinctive bacteriostasis and anti-bacterial bio-agent called 'bamboo kun', which paved the way for health care and medical textile applications, as well as catering the niche of luxury textiles. Moreover, reinforcement with bamboo fibres in composites has been generally found to considerably improve their mechanical and thermal properties [33, 34].

3.5 Banana Fibre

Banana fibre is a lignocellulosic bast fibre extracted from the banana pseudostem (*Musa sapientum*) [35]. Banana fibres are employed in various textile and paper industries owing to its tensile strength and biodegradability aspects. The banana pseudostem undergoes various steps to get banana fibres or fabric including pseudostem harvesting and mechanical extraction of fibres, followed by washing, drying, and subsequent processing of fibres into yarn or fabric [36]. It is grown in 135 countries, mostly in production by India, China, Philippines, Colombia, Indonesia, Ecuador, and Brazil, finding the most ideal conditions for growth in tropical climates. In 2017, global banana production reached 113.9 million tonnes, while it is assumed that about 1 kg of fibre comes out of 37 kg of stems. Bamboo fibres are typically composed of cellulose, hemicellulose, lignin, pectin, and ash in proportions of 60–65%, 6–19%, 5–10%, and 1–3% respectively [35].

3.6 Sustainable Silk

The textile industry is paying more attention to sustainable silk production as a means of minimizing its environmental impact while preserving the luxurious attributes of silk. The conventional production of silk, primarily sourced from the cocoons of the *Bombyx mori* silkworm, raises significant ethical and environmental concerns. These include the heavy reliance on electricity, insecticides, and water, as well as the ethical problems raised during the killing of silkworms for harvesting. Sustainable silk uses a range of cutting-edge techniques and technology to try and solve these issues [36]. Hence, with the sustainability aspect, researchers have thrown light on organic silk farming, Ahimsa silk, silk waste utilization, as well as employing biotechnology in silk production.

(a) **Organic silk farming**: In this method, mulberry trees, the primary food source for silkworms, are cultivated organically, and silk is produced without the use of harmful chemicals. Researchers have explored that organically produced silk has a smooth and luxurious texture similar to conventional silk, as well as adding an eco-friendly badge to it [36, 37].
(b) **Ahimsa silk (Peace silk)**: This type of silk is produced without harming or killing the silkworm, allowing them to complete their natural life cycle. In other words, silkworms are let to mature and come out from their cocoons naturally before the silk is harvested [36]. While getting an eco-friendly tag, this fibre fails to get a good yield, natural look, and mechanical properties than the conventional silk, as well making it a slightly rough textured fibre.
(c) **Utilization of silk waste or vintage silk**: This kind of silk can be regarded as recycled or upcycled silk based on the value adorned to the new product from the silk waste. The waste silk materials from the industries are collected, processed, and spun into new yarn to fabricate a new product with a higher value than the original (upcycling) or lower (recycling or downcycling) [36]. This adds unique textures and patterns to luxury textile products while availing the sustainable tag.
(d) **Wild silk**: This is another kind of silk, harvested from the silkworm living in the forest or wild. Cocoons are collected from the forest after the moths have come out from the cocoon, naturally, producing a coarser texture and natural variation in the fibre colour [36].
(e) **Silk production though biotechnology**: The development of silk derived from genetically modified organisms or microbial fermentation is made possible through technological advancements. For instance, Huang et al. [38] worked on fabricating silk-based materials such as films, sponges, electro-spun mats, and hydrogels by employing recombinant DNA technology. Moreover, AMSilk, a German-based company, spider silk proteins using recombinant bacteria, which may subsequently be spun into fibres, pulverized into powders, or fabricated into a variety of otherwise made into a variety of other shapes [39, 40]. Similarly, Spintex, a company from the Oxford Silk Group at Oxford University, focused on developing a sustainable silk fibre mimicking the 'spider silk' by employing spider inspired spinning process. These fibres are well suited for the production of luxury textile goods that are sustainable [39, 41].

3.7 *Hanji Yarn*

Hanji yarn, made from the cellulosic fibres of the inner bark of mulberry paper (IBMP) trees using techniques derived from traditional Korean paper-making, is a notable sustainable material in luxury textiles. The IBMP pulp fibres are distinguished by their log length and narrow width, offering superior tensile strength compared to yarns with shorter fibres. With an average fibre length of 9.37 mm, second most to hemp at 14.46 mm, and an average width of 0.027 mm, IBMP fibres are classified as highly durable. Hanji yarn also offers inherent antimicrobial properties, excellent moisture absorption, effective wicking, and strong UV radiation-absorbing capabilities. These unique characteristics, derived from the innovative raw materials and production processes, enhance both the aesthetics and functionality of textiles. Highlighting its versatile and eco-friendly qualities, Hyundai Dymo, a subsidiary of Hyundai Motors, unveiled its conceptual car seat covers made from Hanji fabric in 2014. This initiative showcases the potential of Hanji to transform luxury textile applications across diverse product categories [42].

3.8 *Pinatex*

Pinatex is an advanced and non-woven textile crafted from pineapple leaf fibres and is related to biomass. It is produced by harvesting leaves after pineapples have been picked, later extracting the fibres from them, and then processing them into a non-woven fabric [22, 43]. Figure 3 illustrates the various stages of the Pinatex fibre life cycle from raw material extraction to end applications.

3.9 *Mushroom Leather from Mycelium*

Mushroom leather is made up of mycelium, which is the root structure of mushrooms [44]. It is growing more and more well-liked because of its sustainability and other distinctive properties; unlike cowhide leather, it is vegan [45]. The material can be regarded as a sustainable substitute for traditional leather, in luxury textile applications. This leather is lightweight and flexible and can be used to fabricate products ranging from fashion to furniture. Moreover, it is biodegradable and compostable since it is made using non-toxic chemicals and agroforestry waste [23, 44]. The growth and development of the mushroom mycelium life cycle begin with the

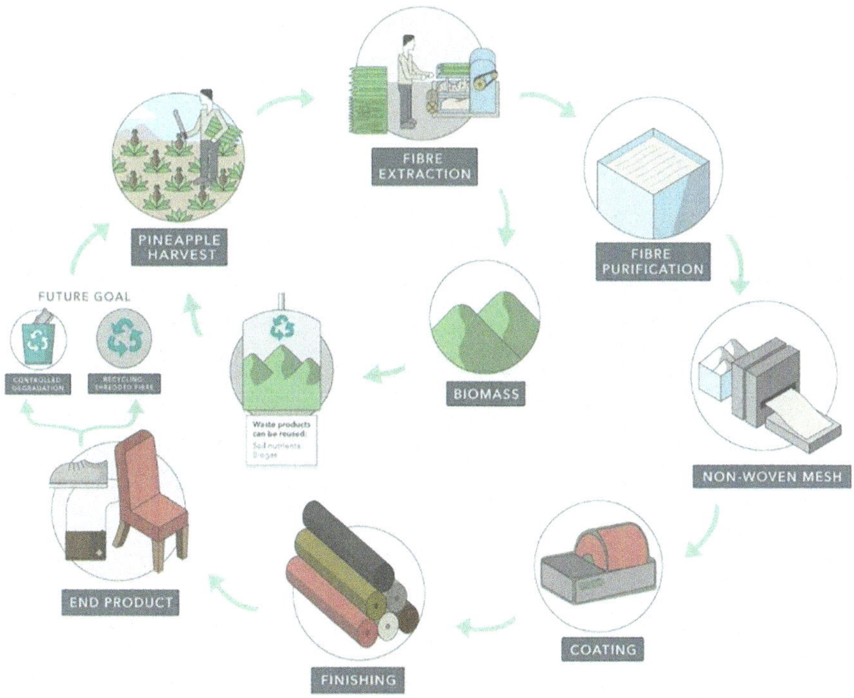

Fig. 3 Production of Pinatex textile from pineapple plant [43]

cultivation of fungal spores, which develop into a vast network of mycelium fibres (Fig. 4). This mycelium is then grown and developed for use in sustainable material uses for products or packaging solutions.

3.10 Algae

Algae are fascinating microorganisms that can be engineered for specific textile use. Algae is a versatile and abundant raw material used in the textile industry for producing fabrics, clothing, and footwear materials. Algal fibres, like *Celp*, are being developed to have antibacterial qualities in addition to improving comfort [46]. It can be harvested daily, hence being very productive, while usual crops, like corn, take two seasons in a year. Companies are researching algae using algae for ink and algal dye production creates a carbon-negative and renewable process that reduces carbon emissions by 200% in comparison with petroleum-based inks and dyes [47, 48].

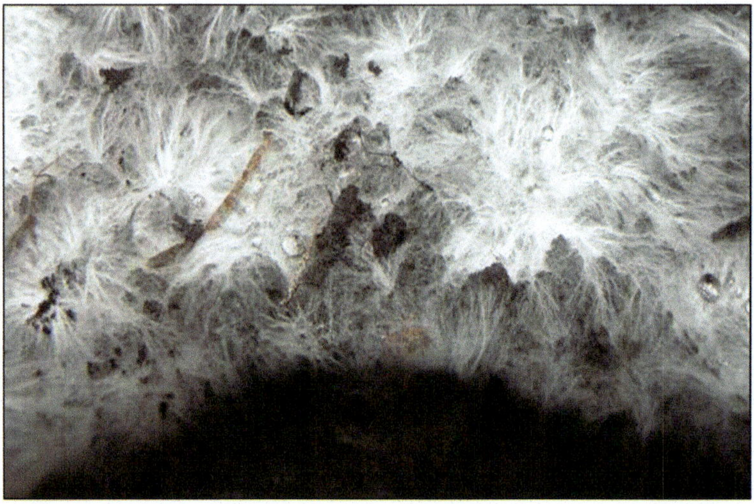

Fig. 4 Mushroom mycelium fibres [44]

3.11 Milk Fibre

Milk fibre, commonly referred to as casein fibre, is a novel fibre with enormous promise for luxurious textiles because of its versatile and environmentally friendly characteristics. As milk fibre is made from renewable milk protein, it is a great material for luxurious textiles since it has a silk-like smooth, and light texture. Milk fibre is a biodegradable fibre derived from renewable resources, which is in line with the luxury fashion industry's rising emphasis on sustainability. The process of making milk fibre entails taking the casein proteins out of skimmed milk and using a mix of mechanical and chemical techniques to turn them into fibres. This method uses less water than other natural fibre manufacturing methods, such as cotton, and makes use of leftover milk products that are unfit for human use, saving food waste and enhancing the value of by-products. To create wonderful fabrics for elite clothing, milk fibre can be combined with other natural fibres such as silk, cashmere, wool, or cotton in luxury textiles [25, 49–52, 53].

3.12 Viscose Rayon

Viscose rayon, the first synthetic cellulosic fibre, was patented in France in 1884, with industrial production starting in 1910. The manufacturing process involves changing raw cellulose first into cellulose xanthate and then dissolving it to obtain a viscose solution, which after extruding through fine nozzles enters an acid-coagulating bath where the cellulose regenerates into filaments. The first and main step in the viscose rayon wet spinning process is forcing the degassed and filtered viscose solution through a spinneret into an acid coagulation bath that contains sulphuric acid among other chemicals [26].

3.13 Lyocell

Lyocell fibres, derived from wood pulp, are recognized for their ecological benefits and suitability as a sustainable alternative to conventional synthetic fibres. Sourced from fast-growing, certified woods such as oak, birch, eucalyptus, bamboo, and beech, lyocell is produced in a closed-loop process where over 99% of solvents are recycled, minimizing environmental impact. The fibres are biodegradable and compostable, decomposing within months, and they are highly valued for their breathability and moisture-wicking properties [54].

Lyocell fibres, such as TENCEL™ from Lenzing, are becoming increasingly popular in luxury fashion due to their eco-friendly closed-loop production process, which recycles 99.5% of water and solvents. This sustainable method along with the traceability of the fibres to managed forests makes lyocell an attractive choice for high-end brands committed to quality and environmentally responsible [27, 55].

3.14 Recycled Cotton

Recycled cotton, also called regenerated or reclaimed cotton (or shoddy), is created by converting cotton fabric back into fibre for reuse in textiles. Sources for recycling include pre-consumer waste, such as scraps from yarn and fabric production, and post-consumer waste, such as used clothing and household items. Pre-consumer waste, particularly cutting scraps, is the primary source of recycled cotton. In contrast, sorting post-consumer waste is more labour-intensive due to varying colours and fabric blends [56, 57]. Recycled cotton can fit in fabricating various ranges of luxury textile goods from luxury upholstery to garment wears.

3.15 Recycled Polyester

The fashion industry has a long journey toward full sustainability, but some brands are making strides to lessen their environmental impact by incorporating recycled materials. Recycled polyester fibres, for example, are being produced from discarded Polyethylene Terephthalate (PET) bottles and polyester-based garments. REPREVE(r) leads in this, converting billions of recycled plastic bottles (chiefly of PET) into high-performance yarn, which is used in products from apparel and

Fig. 5 Productions of polyester garment from PET bottles [28]

accessories to shoes and tents. These would include sorting, cleaning, and processing PET bottles into flakes, which afterward turn into new fibres for the making of fabrics [28, 58, 59]. The steps in the production of recycled plastic or polyester textiles from PET bottles are depicted in Fig. 5.

3.16 Recycled Nylon

Recycled nylon is made from used nylon wastes, such as discarded products and leftover by-products during manufacturing. It also represents an environmentally friendly substitute for virgin nylon, whose raw material comes from chemicals taken from petroleum. Manufacturing recycled nylon involves breaking down nylon wastes, which could be in the forms of discarded fishing nets and plastic scraps, into their chemical components, and recombining them into nylon sheets. Recycled nylon reduces pollution and waste in landfills, hence contributing to a lower carbon footprint. The application of recycled nylon can be found in making bags, clothes, and accessories. It is suitable for manufacturing fishing nets and helps in reducing ocean waste by supporting environmental sustainability with minimal waste and pollution [60, 61].

3.17 Microfibre

A synthetic product known as microfibre is manmade and weighs less than one-fifth the diameter of human hair. It has two important structural constituents: polyester, which provides the body, and polyamide, which contributes to absorption and density. Microfibres work very well for a variety of applications, including luxury clothes, upholstery, towels, and cleaning supplies because of their ability to collect tiny dust and dirt particles [62].

3.18 Polyethylene Furanoate

Polyethylene Furanoate (PEF) is a biopolymer synthesized from bio-based feedstock, featuring fibre properties without compromise. It is produced by combining furanoate monomer, 2,5-furandicaboxylic acid, with bio-based ethylene glycol to produce a product. Owing to its potential for environmentally friendly alternatives, PEF fibres found use in a variety of industries, including home furnishings, clothing, and fabric manufacture. The applicability also covers industrial fibres, infant products like diapers, and disposable consumer goods that are 100% bio-based polymers [63, 64].

3.19 Polyhydroxyalkanoates

Polyhydroxyalkanoates (PHAs) are 100% bio-based polymers, making them fully biodegradable and a sustainable option for various applications. Their versatility begins from being produced from a wide range of biomaterials, and even carbon dioxide, contributing to their environmental benefits. In luxury textiles, PHAs offer a sustainable alternative, particularly suited to environments with high microbial activity, as they degrade rapidly under such conditions, augmenting their appeal for eco-conscious brands [63].

4 Sustainable Chemicals in Luxury Textile Production

Textile production, particularly wet processing, depends extensively on various chemicals throughout different stages, which vary based on the material type and intended use [65, 66]. Although these chemicals are vital for achieving specific textile qualities such as colourfastness, wrinkle resistance, and softness, as well as other processes, they can also present considerable environmental and health hazards [65, 67, 68]. As a result, the use of sustainable chemicals in luxury textile manufacturing is becoming increasingly important. The various stages of textile wet processing in which chemicals are involved in textile processing are shown in Fig. 6.

4.1 Sustainable Chemicals in Textile Pre-treatment Stage

Textile pre-treatment is an important step in textile processing where many chemicals and auxiliaries are used. The following section describes the different pre-treatment processes and the chemicals involved and suggests sustainable alternatives.

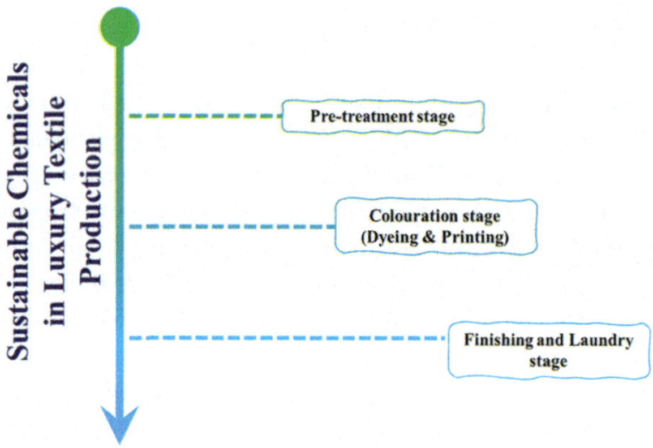

Fig. 6 Stages in the application of sustainable chemicals in luxury textile production [65]

4.1.1 Desizing

This process involves the removal of affixed sizing agents (starch, polyvinyl alcohol, or guar gum) through acid or enzyme-assisted hydrolysis since these agents can obstruct the penetration of dyes and chemicals into the fibre. To achieve a more sustainable process, desizing is often performed by hydrolysing the sizing materials using amylase enzymes [68, 69]. Amylase enzymes can be utilized for desizing processes at lower temperatures ranging from 30 to 60 °C, with an optimal pH between 5.5 and 6.5 [70].

4.1.2 Scouring

Scouring eliminates both adhered and added fatty substances from textile materials through a saponification process. Impurities such as pectin, fat, hemicelluloses, oil, minerals, natural colours, and wax are commonly found in natural fibres [71]. These substances are mainly acidic and may be efficiently eliminated in hot, alkaline environments [68]. For bio-scouring, cellulase and pectinase are usually combined. By digesting the pectin, pectinase tears down the structure of the cotton cuticle and cuts the link between the cuticle and the main body of the cotton fibre. Meanwhile, cellulase works by degrading the primary wall cellulose located just beneath the cuticle, further aiding in the removal of the cuticle structure [70]. Recently, the use of lipase-based scouring, as well as combined scouring and bleaching processes, has gained favour for their economic efficiency and sustainability [69].

4.1.3 Bleaching

Using reducing or oxidizing chemicals, bleaching removes the fibres' inherent pigments. In the fabrication of luxury textiles, hydrogen peroxide-based bleaching is preferred due to its sustainability and longevity. To save water and energy, a combined scouring and bleaching method is frequently employed for darker hues. Additionally, catalase enzymes are employed to break down residual hydrogen peroxide after the bleaching process, further enhancing the sustainability of the chemical treatment [68, 69]. Additionally, laccases can be employed to bleach wool fabrics and indigo-dyed denim, offering an eco-friendly alternative to traditional chemical bleaching methods [70, 72].

4.1.4 Degumming of Silk

Degumming, which involves removing the gummy sericin (20–25%) from silk cocoons, is typically achieved using a hot alkaline solution. This process enhances the lustre, colour, texture, and hand feel of silk materials. For a more sustainable approach in luxury textile production, sericinase enzymes can be used to degum silk substrates, offering an eco-friendly alternative to traditional methods [68].

4.1.5 Mercerization

Mercerization, traditionally applied to cellulosic textiles using 250 g/L of NaOH (sodium hydroxide), enhances fabric reactivity and lustre by partially converting cellulose into soda cellulose, which improves dye uptake and fixation, particularly in light-to-dark shades. To align with sustainable practices in luxury textile production, the focus is on reducing the sodium hydroxide (NaOH) concentration while maintaining the intended mercerized outcome [69].

4.2 Sustainable Chemicals in Textile Colouration (Dyeing and Printing) Stage

After the preparatory stages, the next phase in textile treatment involving chemicals is the colouration stage. This process includes the use of fixing agents, chemical mordants, carrier agents, synthetic dyes, pigments, and more. With sustainability in mind, the following section highlights various eco-friendly alternatives used in the textile colouration process.

4.2.1 Natural Dyes

These are the ancient ingredients most commonly used to render pleasing colours on to the textiles. However, the invention of synthetic dye and the chemical revolution replace natural dyes in the industry by addressing the shortcomings. The textile industry is now, albeit rather late, compelled to utilize natural dyes for textile colouring, primarily in the luxury market due to their expensive nature [73].

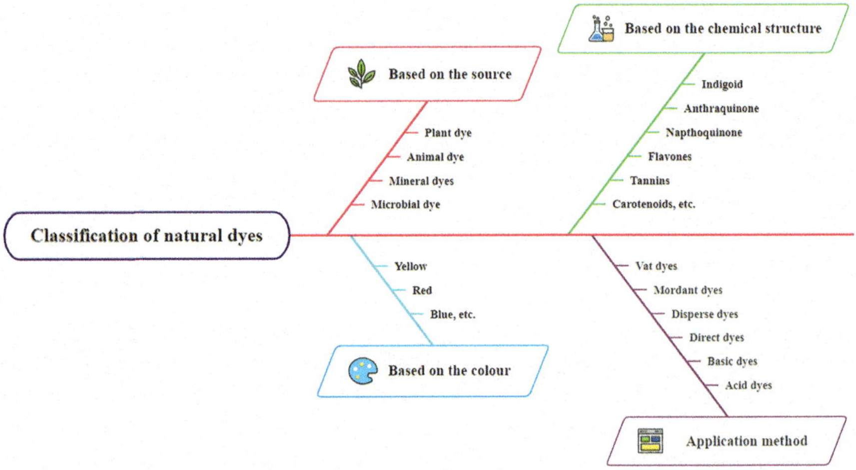

Fig. 7 Various classification of natural dyes [74]

Natural dyes are colourants derived from natural sources, such as plants (vegetables), animals, or minerals. These dyes are sustainable and eco-friendly owing to biodegradable, renewable, and less-toxic characteristics compared to conventional synthetic dyes. More importantly, the elegance and value of textiles are greatly increased when natural dyes are used. As shown in Fig. 7, naturally derived dyes can be classified into a number of subclasses based on the source (plant, animal, mineral, and microbial), chemical structure (indigoid, anthraquinone, napthoquinone, etc.), colour (red, blue, yellow, etc.), and application method (acid, basic, direct, disperse, mordant, and vat dyes) [74].

Furthermore, vegetable or plant dyes are further categorized into plants dyes (madder and indigo), animals (cochineal insects and certain species of molluscs), minerals (ochre and clay), and microbes (bacteria and fungi), with plant-based sources being the most prevalent. Different parts of a plant, such as leaves, roots, twigs, stems, heartwood, and bark, can all be utilized to produce various shades of colours on the textile substrate [75].

Although natural dyeing of textiles is an age-old technique known for its sustainability, recent advancements in the textile industry have further raised its value by improving the efficiency and yield of these dyes. Pizzica et al. described the advancements in natural dye extraction including ultrasound- and microwave-assisted extraction, aqueous extraction, solvent extraction, supercritical fluids, acid or alkali extraction, and enzymatic extraction and fermentation [74].

Natural dyes are abundantly existing in the environment, and numerous sources have already been recognized by researchers. These dyes furnish distinctive aesthetic looks and eco-friendly solutions, making them useful for producing or improving luxurious textile goods. However, many potential natural dye sources have yet to be discovered. The various natural dyes obtained from vegetable and animal sources are explored in studies by [73–85].

4.2.2 Bio-mordants for Natural Dyes

In order to fasten the colour of fabrics and create a stable dye–metal complex, mordants are traditionally used to treat fabrics so as to boost the poor substantivity and insignificant fastness characteristic of natural dyes. The mechanism of mordant is subjected to the kind of fibre employed: for protein fibres, the dye molecule binds to the mordant, but for cellulose fibres, the mordant remains an insoluble component on the fibre, while the dye molecule binds to it.

Owing to the risks associated with the traditional mordant used in natural dyeing on the environment, the search for eco-friendly alternatives has advanced. Recently, researchers have explored bio-mordants from plants and biomaterials as a sustainable option over conventional metallic mordants [86–88]. These bio-mordants are available in a wide range and their effect chiefly depends on their content and chemical structure. Bio-mordants can be classified as rich in tannins, rich in natural acids, rich in protein, and other bio-materials. The following are some of the commonly available bio-mordants that can be employed in luxury textile dyeing [89]:

Tannin-rich plant mordants: Myrobalan (*Terminalia chebula*), pomegranate rind (*Punica granatum*), oak bark or gall nuts (*Quercus* spp.), cutch (*Acacia catechu*), tea leaves (*Camellia sinensis*), walnut hulls (*Juglans regia*), chestnut (*Castanea* spp.), eucalyptus bark (*Eucalyptus* spp.), pomegranate (*Punica granatum* L.) rind, and Arjun (*Terminalia arjuna*)

Natural acid-based mordants: tamarind (tartaric acid), lemon juice (citric acid (CA)), and vinegar (acetic acid)

Protein-rich materials: Soy milk, casein, lactarane, albumen, and urine

Other bio-mordants: Rhubarb leaves (*Rheum rhabarbarum*), onion skins (*Allium cepa*), mango leaves (*Mangifera indica*), chitosan, green tea, coffee, pine (*Pinus pinea* L.) cone and nutshell, orange peel, ammonium acetate, caffeic acid, white mulberry (*Morus alba* L.) fruit, clays, egg shells, jaggery, lime, and so on

4.2.3 Microbial Dyeing of Textiles

Using microorganisms particularly bacteria and fungi to create dyes is a novel and sustainable method of textile colouration known as microbial dyeing. Due to the environmental problems posed by traditional dyeing processes—which frequently include hazardous chemicals, large water consumption, and substantial waste generation—this approach has gained popularity. Through the employment of microorganisms in producing and extracting natural pigments, the textile industry may both create distinctive and vivid colours while lessening its environmental impact.

The application of biotechnology in textiles offers a promising approach to the commercial production of colourants through microbial fermentation. This method of producing pigments has several advantages, including lower costs, higher yields, easier extraction, and a reliable supply of raw materials without seasonal limitations. For instance, the use of fungal dyes in textile colouration can significantly contribute to sustainable and environmentally friendly practices, reducing the need for natural resources such as plants and animals and helping to prevent deforestation, conserve species, and maintain biodiversity in the ecosystem [90, 91].

The bacteria and fungi are such specific microorganisms that could generate a wide range of pigments useful in the textile colouration processes.

- **Bacterial pigments**: Some bacteria like *Chromobacterium violaceum* produce pigments like violacein, which could lend purplish tones to textiles. Others such as *Pseudomonas putida* could produce yellow bacterial pigments or *pyoverdine* with other colours through metabolic activities [92].
- **Fungal pigments**: Microbial dyes are gaining attention as a sustainable and safe substitute for synthetic dyes. Fungi, in particular, have been identified as promising sources of natural pigments, such as naphthoquinones. The pigments derived from fungi provide a dependable source of organic colourants with enhanced functionality. Chemically, these fungal pigments fall under categories such as carotenoids and polyketides. Several fungi species, including *Streptomyces*, *Paecilomyces*, *Monascus*, *Cordyceps*, and *Serratia*, are known to produce various pigments through fermentation. It has been explicitly reported that certain strains of *Bacillus* species, *Phaffia rhodozyma*, *Monascus purpureus*, *Penicillium atrovenetum*, and *Penicillium herquei* produce yellow, red, and blue pigments [93].

Additionally, certain strains of the Talaromyces genus (formerly known as Penicillium), including *Talaromyces pinophilus*, *T. funiculosus*, and *T. aculeatus*, have been investigated for their potential to produce polyketide azaphilone colours that resemble *Monascus*. Furthermore, species such as *Monascus purpureus* and *Pseudomonas fluorescens* have been utilized for dyeing textiles such as cotton, wool, and silk [94–96].

4.2.4 Natural Indigo Digital Textile Printing Ink

Natural indigo DTP (digital textile printing) ink represents a significant advancement in sustainable chemical use within luxury textile production. Derived from the indigo plant, this eco-friendly alternative to conventional synthetic dyes combines the time-honoured tradition of natural dyeing with cutting-edge digital printing technology. Unlike synthetic dyes, which often involve toxic chemicals and contribute to environmental degradation, natural indigo is biodegradable and poses minimal risks to both human health and the ecosystem. The extraction process of indigo, which emphasizes minimal waste and maintains the dye's integrity, further underscores its sustainability. As the luxury textile industry increasingly shifts towards environmentally responsible practices, the use of natural indigo DTP ink not only aligns with this trend but also enhances the marketability of products by appealing to eco-conscious consumers. Moreover, the unique aesthetic properties of natural indigo, with its ability to produce varied hues depending on the fabric and dyeing method, offer designers the chance to create distinctive, luxurious textiles that marry tradition with innovation [97, 98].

4.3 Sustainable Chemicals in the Textile Finishing Stage

The growing emphasis on sustainability, driven by heightened consumer awareness, stricter regulations, and proactive eco-initiatives, has incited a significant shift within the industry. There is a growing preference for bio-based, natural, and eco-friendly materials and products over conventional alternatives, aiming to mitigate environmental impact and foster a more sustainable future. The following contents discuss the sustainable chemicals employed in various textile finishing stages.

4.3.1 Sustainable Water and Soil Repellent Treatment on Luxury Textiles

The demand for luxurious, high-performance textiles that can withstand the elements while remaining environmentally conscious is on the rise. Sustainable water and soil-repellent treatments offer a way to protect delicate fabrics from stains and moisture damage without compromising on ethical and ecological values. Moreover, Table 2 describes numerous sustainable alternatives for soil and water-repellent finishes for luxury textile applications.

4.3.2 Wrinkle-Resistant Luxury Textiles

Wrinkle-resistant finishes are essential for luxury textiles such as cotton shirting and suiting to maintain their refined appearance. Traditionally, formaldehyde-based resins such as dimethylol dihydroxy ethylene urea have been used to provide

Table 2 Eco-friendly alternatives for water and soil repellent finish

Eco-friendly alternative	Description	References
Fluorocarbon-free technologies	These include silicone-based polymers, hydrocarbon-based polymers, and dendrimer-based polymers. These provide effective repellency while minimizing environmental impact	[99]
Plant-based oils and waxes	These natural substances, such as soybean oil, carnauba wax, and beeswax, offer renewable and biodegradable alternatives to conventional chemicals	[100–102]
Sol-gel technology	This versatile method allows for the creation of thin, durable, and breathable coatings with excellent repellency properties. Sol-gel coating using TiO_2 particles, 3-aminopropyltriethoxy silane (APTES) and hexadecyltrimethoxysilane (HDTMS), SiO_2 nanoparticles, polydimethylsiloxane (PDMS)-silica nano particles, Al_2O_3, ZnO, ZrO_2 and TiO_2 nanoparticles and so on can also be used to furnish water repellent finish on various textile materials	[66, 103–105]
Plasma treatment	This dry process modifies the surface of textiles, creating low surface energy and thus enhancing repellency. Plasma treatment is considered environmentally friendly as it doesn't involve the use of water or chemicals	[66, 106]
Nanotechnology	The use of nanoparticles, like silica or zinc oxide, can create highly effective soil repellent finishes	[107, 108]

durable press properties, but they pose significant health risks, including irritation and cancer, leading to a demand for safer alternatives. Sustainable options, including polycarboxylic acids such as CA, 1,2,3,4-butanetetracarboxylic acid (BTCA), succinic acid, maleic acid, itaconic acid, and malic acid, have been developed, with BTCA gaining commercial success due to its effective wrinkle resistance without compromising fabric quality. Additionally, nanotechnology has emerged as a promising approach, using titanium dioxide (TiO_2) and nano-silica as catalysts to enhance wrinkle resistance by creating cross-linking agents that bond with cellulose in fabrics. These advanced methods offer safer and more sustainable alternatives to traditional chemical treatments while maintaining fabric comfort, making them ideal for use in luxury textiles [67, 109].

4.3.3 Eco-Friendly Flame-Retardant Finish

A flame-retardant (FR) finish is a specialized treatment for luxury textiles designed to reduce and limit fire hazards when these fabrics come into contact with fire sources. The conventional FR chemicals were mainly based on chlorine and bromine compounds or contained nitrogen and phosphorus elements; these proved to be potentially hazardous and toxic to both humans as well as the environment [110]. This paved the way for looking out for sustainable FR agents or chemicals. Table 3 indicates the various bio-based fire retardants and their characteristics. Researchers explored various innovative bio-based FRs from chitin, phytic acid, eggshell, oyster shell powder, and rice husk [111–114]. Additionally, various researchers have synthesized bio-based FRs for sustainable FR treatment on textiles such as phosphorous-containing FRs [115], phosphorous-based FR [116], isosorbide-based polyphosphonates [117], polybasic carboxylic acid [118], bio-based macromolecule ammonium phytic acid [119], and phosphorous/furan-containing FR [120].

Table 3 List of some bio-based fire retardants [110]

Bio-based FR's	Source	Characteristics
Tannic acid	Natural polyphenols- richly accessible in plants	The flame quenching properties along with the superior char formation during combustion
Phytic acid	Plant tissues contain a large amount of phosphorous-containing compounds	Prevents the spreading of flame by forming char
Isosorbide 15	Obtained directly from D sorbital (hydrolysis process of starch)	–
Diphenolic acid 24	Produced by the chemical reaction between levulinic acid and phenol in the presence of hydrochloric acid	–
Lignin	Wood and bark	The superior char formation during combustion
β–Cyclodextrin	A macrocyclic oligosaccharide containing seven subunits of glucose	Smooth decomposition & high yield of chars

4.3.4 Sustainable Moth Proofing for Luxury Woollen Textiles

Moth proofing is essential for preserving the quality and longevity of luxury woollen products, such as premium garments, blankets, and carpets. Wool, a natural fibre high in keratin, is particularly vulnerable to damage from moth larvae, notably from species such as *Tineola bisselliella* (the common clothes moth) and *Anthrenus verbasci* (the varied carpet beetle). These pests can cause extensive damage, leading to unsightly holes and the degradation of valuable textiles, making effective moth proofing crucial. Increasingly, there is a shift towards natural moth repellents, which are seen as more eco-friendly and safer for use in households. Natural substances such as cedarwood, lavender, and rosemary are well-regarded for their moth-repellent properties and can be applied via sachets or sprays to protect woollen products. While these natural solutions are effective, they often require more frequent reapplication than chemical treatments [121, 122]. Advances in textile technology have introduced nano-coatings as a modern solution for moth proofing. For instance, research indicates that nano-sized titanium dioxide (TiO_2) can offer robust protection against moths while preserving the fabric's integrity, presenting a sustainable option for moth-proofing luxury textiles [123].

4.3.5 Cosmeto-Textiles: Sustainable Textile Finishing Using Plant Extracts

Luxury cosmeto-textiles have emerged with integrating traditional knowledge from Ayurveda, cosmetics, as well as textile science, offering benefits such as slimming, moisturizing, perfuming, and promoting overall well-being. Textiles treated with plant extracts rich in minerals, secondary metabolites, co-enzymes, and antioxidants can address issues such as nervousness, weakness, cholesterol levels, mental stress, skin conditions, and cardiac health [67]. For example, *Clevertex* invented wellness bath towels that have been enhanced with ZnO and seaweed extract that are integrated right into the fibres. These towels provide comfort and stress relief while maintaining their functional properties, such as antimicrobial and liquid absorbency, even after repeated washing and wearing [124]. Similarly, *KHEONI* Company has developed a therapeutic luxury eye pillow designed to provide instant stress and anxiety relief, alleviate headaches, and improve sleep quality. This product is finished with extracts from dried ginger, lemongrass, tamarind leaves, and pure essential oils [125].

4.3.6 Sustainable Fragrance Textiles

In the realm of luxury textiles, the integration of fragrance and aroma treatments has become a notable feature, offering wearers an energetic, refreshing, and hygienic experience. Techniques such as microencapsulation, nano-encapsulation, or direct finishing enable the embedding of plant-derived aromas into the textile substrate, providing a sensory appeal that enhances the luxury experience. For instance, Bombay Dyeing's fragrance-based bed linen collection offers a soothing and refreshing ambiance, featuring soft cotton satin with a high fabric count and natural scents of jasmine, rose water, and lavender. These textiles not only create a pleasant atmosphere but also help reduce mental stress, promoting comfortable sleep. The fragrance is slowly released over time as the microcapsules embedded in the fabric burst due to pressure or friction, ensuring long-lasting freshness [126]. In line with sustainable practices, Singh et al. [127] explored an innovative approach to luxury textile finishing by applying lavender essential oil, β-cyclodextrin, and chitosan citrate to cotton fabrics. This sustainable method enhances the aesthetic appeal and adds antimicrobial properties, addressing both consumer preferences and health concerns. By using natural ingredients and biodegradable polymers, this approach fits nicely with the rising demand for luxury fabrics that use eco-friendly processes, underscoring the environmental compatibility of materials sourced from renewable resources.

4.3.7 Enzymes for Luxurious Textile Finishing

Enzymes have become pivotal in transforming textile processing, particularly for luxury fabrics, by proposing a sustainable and efficient substitute for conventional methods. These natural proteins enable the production of high-quality textiles while significantly reducing environmental impact. Enzymes are integral at various stages of luxury textile finishing, including desizing, scouring, bio-polishing, and softening. They work by selectively targeting specific components within the fibres, leading to improved durability, vibrant colour retention, and a softer feel. A major advantage of enzyme use in luxury textiles is their specificity. Each enzyme is engineered to act on a particular substrate, allowing for precise modifications without compromising the fabric's overall integrity. For example, cellulases target cellulose in plant-based fibres such as cotton and linen, enhancing the texture and appearance of the fabric without affecting other elements. Various enzymes employed for luxurious textile finishing are summarized in Table 4, highlighting their specific roles and contributions to sustainable textile production [68, 128–130].

Table 4 Various enzymes employed for luxurious textile finishing [68, 128]

Enzymes	Effect	Details
Cellulases	Bio-polishing	Removes surface fuzz and pills, leading to a smoother, softer hand feel, and improved lustre
	Anti-pilling	Prevents the formation of pills, enhancing fabric durability and appearance
	Colour brightening	Enhances colour vibrancy by removing surface impurities that can dull colours

(continued)

Table 4 (continued)

Enzymes	Effect	Details
Pectinases	Bio-scouring	Removes natural impurities like pectin from cotton and other plant-based fibres, improving absorbency and preparing the fabric for further processing
	Softening	Contributes to a softer hand feel by modifying the fibre surface
Proteases	Bio-polishing	Improves lustre, smoothness, and drape of protein-based fibres (wool, silk)
	Anti-felting of wool	Reduces wool shrinkage and felting during washing
	Degumming of silk	Removes sericin, the natural gum that coats silk fibres, for a softer and more lustrous finish
Laccase	Bio-bleaching	An environmentally friendly alternative to chlorine-based bleaching, leading to a gentler and more sustainable process
	Colour modification	Can be used to modify and enhance the colour of certain dyes, expanding design possibilities
Amylases	Desizing	Efficiently removes starch-based sizes used in weaving, improving fabric absorbency and preparing it for further processing

4.3.8 Bio-based/Biodegradable Binders

Binders are chemical compounds used to bind other materials to the fabric's surface or to the fibres within its structure. They form a film or matrix that encapsulates these materials, boosting the fabric's durability, stability, and imparting specific functional properties to the finished product. These binders are employed in various textile finishing processes such as pigment printing, coating and laminating, functional finishing, non-woven fabric production, and technical textile finishing. As a replacement for conventional chemical or petro-chemical-based resins, researchers have developed bio-based or biodegradable binders for textile finishing. These binders are generally obtained through extraction, synthesis from renewable resources, or microbial action [16].

Bio-based binders can be of two groups, such as (i) drop-in biopolymers (these are made out of renewable resources but identical to the current petroleum-based polymers) and (ii) new bio-based polymer chemistry (like polylactic acid). Table 5 indicates the various bio-based/biodegradable binders used in textile finishing processes.

Table 5 Biobased/biodegradable binders used in textile finishing processes [16]

Biobased/biodegradable binders	Company & brand name	Characteristics
Drop-in biopolymer		
Biobased thermoplastic PU	Merquinsa/Pearlthane® & Pearlbond ECO®	Biobased content of 20–90%
Biobased thermoplastic copolyamide hotmelt Platamid®	Arkema/Platamid®	Based on castor oil. Biobased content up to 100%
EcoSphere® biolatex binder dispersions	EcoSphere®	Based on starch derived from plants such as corn, potatoes, and tapioca
		Used as a replacement for petroleum-based styrene butadiene latex
EcoMer®	EcoSynthetix®	Based on waterborne sugar-acrylic adhesives and resins
Thermoplastic polyamide, pine-based, adhesive resins	Arizona chemicals/ UNI-REZ™	–
New biopolymer chemistries		
OC-biobinder™	OrganoClick	Derived from carbohydrates, improves the mechanical properties of textiles such as tensile strength, dimensional stability, and stiffness
Biodegradable soy-based polymers	DuPont/SoBind™ balance	Industrial coating applications
Amorphous PLA (hotmelt and water-based dispersions)	Toyobo	Coating applications, mainly in packaging
		Adding glycerol and fatty acids with PLA enhances flexibility

5 Studies Related to Sustainable Fibres, Polymers, and Chemicals in Textiles

The following section describes various studies engaged by researchers, industries, or brands in successfully employing sustainable fibres, polymers, and chemicals in luxury textile production.

5.1 Ecofoot*'s Hybrid Pigments*

For luxury textile dyeing, *Ecofoot* has developed novel hybrid pigments that allow for interactions with cellulose fibres at temperatures as low as 25 °C. These pigments are created by chemically bonding a dye to a polymer particle. This system eliminates the need for salt, a common but environmentally challenging component in conventional dyeing processes. *Ecofoot*'s technologies offer an ecological, energy-efficient, and water-saving solution, meeting the demand for sustainable luxury textile dyeing.

***Ecofoot*-Indigo**: This is employed in denim dyeing and prevents traditional toxic-reducing agents that convert indigo into a water-soluble form, preventing harmful sulphite and sulphate discharge into wastewater.

***Ecofoot*'s auxiliaries**: These prevent dye hydrolysis, reducing the need for harsh washing-off procedures and saving over 50% of water in the dyeing process [131].

5.2 Recycrom™ Readytodye *by Officina39*

Recycrom™ readytodye is a sustainable dye fabricated from recycled fibres, clothing, and textile scraps, developed by an Italian company, Officina39 (Fig. 8). This dye stuff contains a minimum of 65% textile waste from pre- and post-consumer waste that is converted into an ultra-fine powder that acts as a pigment dye for various fabrics, including cotton, wool, nylon, and other natural fibres. *Recycrom™ readytodye* can be applied using multiple methods such as dipping, spraying, exhaustion dyeing, screen printing, as well as coating. *Recycrom™ readytodye* dye stuff is applied as a suspension, in contrast to traditional dyes, which are applied as chemical solutions. This minimizes the dyeing process's negative environmental effects by making water filtering simple.

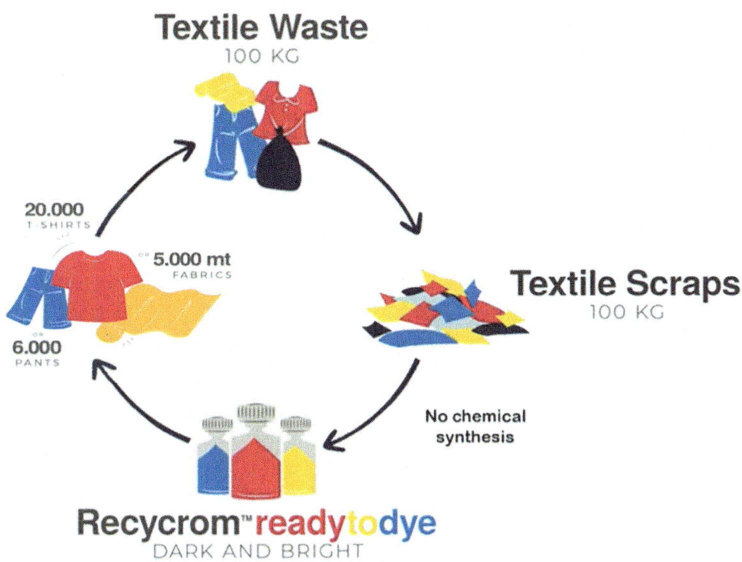

Fig. 8 The recycled dyestuff technology by Officina39 [132]

5.3 Engineered Microorganisms by Colorifix

Colorifix is an innovative start-up that produces, deposits, and fixes pigments onto fabrics using a biological method. Employing a synthetic biological approach, *Colorifix* utilizes bacteria to colour fabrics, potentially reducing water usage by up to ten times compared to traditional dyeing methods. The method uses a carbon source solution to adhere bacteria that produce colour directly onto the fabric. The dye is then deposited and fixed in a single heating cycle by lysing the bacteria. This method eliminates the need for conventional dye extraction processes, which typically use organic solvents, fixing agents, or reducing agents containing harmful organic compounds, making it a more sustainable and eco-friendlier alternative for textile dyeing [131].

5.4 Sustainable Dyeing Technologies by Huntsman Textile Effects

AVITERA® SE (a line of polyreactive dyes) and Univadine E3-3D (a dyeing auxiliary) were launched by Huntsman Textile Effects for sustainable dyeing of textile substrates.

AVITERA® SE polyreactive dyes:

- It is designed specifically for cotton with enhanced bonding efficiency compared to conventional reactive dyes.
- It uses tri-functional chemical reactivity, which has three reactive groups in each dye molecule rather than the typical two, to achieve a high rate of reaction and fixation with cellulosic fibres.
- It minimizes the presence of unfixed dye, reducing energy and water consumption by up to 50%.
- It uses up to 20% less salt than traditional dyeing processes.

Univadine E3-3D diffusion accelerant:

- It is developed as a dyeing auxiliary to enhance dye diffusion into polyester fibres, especially microfibres.
- It achieves high-performance dyeing without the use of toxic chemicals.
- It meets current and future industry sustainability criteria [133].

5.5 GENTLE POWER BLEACH™: Innovative Low-Temperature Textile Bleaching

GENTLE POWER BLEACH™ is an advanced peroxide bleaching system that uses enzyme innovation for effective low-temperature bleaching at neutral pH. It enables bleaching at 65 °C (149 °F) and neutral pH, reducing water and energy consumption by up to 40% during pre-treatment. The following are the benefits of this technology:

- It minimizes water and energy use by about 50% and enhances productivity by 10%, while achieving high-quality full white levels, particularly on regenerated cellulosic fibres.
- It also avoids the use of harsh chemicals, lowering the risk of crease marks on sensitive fabrics such as cotton and Lycra.
- It protects delicate and high-end fibres such as viscose, modal, Tencel, and their blends from damage.
- It minimizes fabric weight loss, enhances colour yield, and produces brighter shades [133].

5.6 Leading Luxury Brands and Sustainability Initiatives

- *Gucci: Focus on reducing carbon emissions and using sustainable materials such as organic cotton and Mylo*: Marco Bizzari and Designer Alessandro Michele has implemented several sustainable initiatives for the brand Gucci. "Gucci Equilibrium" was launched to consolidate the environmental and social efforts to foster sustainability and innovative eco materials to tackle global pollution. Gucci announced the use of Mylo in 2018, a leather made from mycelium, the root structure of mushrooms, an alternative material for traditional leather. Gucci has committed to reducing its carbon footprint by 50% by 2025 and 95% organic cotton has been incorporated in its latest collections [134].
- *Stella McCartney: Pioneering cruelty-free fashion, using sustainable and innovative materials like vegan leather and bio-based alternatives.* Stella McCartney continues to be the pioneer in the application of sustainable materials in the family of luxury fashion. Ever since 2001, The brand has been avoiding the use of leather, fur, and feathers. Instead, the company uses alternative materials such as vegan leather, made from polyurethane and other recycled materials and one such is Econyl. The brand launched an Econyl collection in 2019; Econyl is a regenerated nylon made from plastic waste, for which every 10,000 tonnes of recycled nylon saves 70,000 barrels of oil [134].
- *Burberry: Carbon-neutral strategy in runway shows and using recycled materials in its collections.* Burberry has adopted several sustainable practices, including the use of recycled and organic materials. In 2020, Burberry collaborated with a company "Evrnu" known for developing new textile fibres from old clothing waste materials, launched a limited-edition capsule fashion collection using recycled cotton and polyester made from recycled plastic bottles. The company has reported that these initiatives have reduced greenhouse gas emissions by 43% since 2016 and achieved products of sustainable components by 68% [134].
- *Hermès: Bets on Mushroom-Based 'Leather' powerhouse is testing alternative materials in a break with tradition.* Hermès is known for its heritage and traditional craftsmanship of leather bags. Hermes collaborated with MycoWorks, a Californian start-up company that manufactures a new mushroom-based alternative to leather in 2021. Hermès is breaking with tradition of animal leather bags

with new initiatives of Vegan Leathers that are more environmentally and socially sustainable [135].

5.7 *Successful Implementations of Sustainable Polymers and Fibres*

- *Econyl by Prada*: The traditional production of nylon is very resource-intensive and releases a significant amount of carbon emissions, whereas Econyl is a regenerated nylon created from nylon waste materials, offer a circular solution allowing for continuous recycling without losing its quality. Prada used such nylon waste materials such as fishing nets and fabric scraps for their collection by collaborating with Italian companies such as Aquafil and Healthy Seas. This collaboration highlights their commitment to sustainability by transforming waste materials of fishing gear and other fabric scraps into high-quality recycled nylon called "Econyl." Incorporating ECONYL® into their collections helps to replace all virgin nylon with this alternative [136].
- *Adidas and Parley for the Oceans: Collaboration for footwear made from ocean plastic waste*. Similar to Prada's collaborations, Adidas partnered with Parley, announcing their collaboration for developing an innovative shoe made from recycled plastics from the ocean at a United Nations conference. The illegal deep-sea gillnets were cleaned and repurposed into yarn and were used in the upper layer of the shoe. This initiative marked a significant step for Adidas to combat plastic waste. This partnership has not only been achieved by reducing the usage of virgin polyester by 50% in the year 2020 but also made 30 million pairs of shoes by developing a yarn in collaboration with Parley Ocean Plastic named "Primeblue" [137].
- *Sustainable Viscose and Forest Conservation by Stella McCartney*: Over 200 million trees are cut every year to create fabric with an increasing portion of viscose generated from endangered forests, leading to deforestation and environmental degradation. Stella McCartney, a luxurious fashion brand, addresses these issues by collaborating with Canopy and sourcing viscose from FSC-certified suppliers to ensure a transparent and sustainable supply chain. Innovations like Evrnu and NuCycl made by their suppliers, ENKA and LENZING aims to replace virgin wood pulp with these recycled materials [138].

6 Conclusion

The luxury textile industry is at a pivotal moment where embracing sustainable practices has become essential rather than optional. By incorporating sustainable polymers, eco-friendly fibres, and greener chemical processes into its production chain, the industry can transform its operations and significantly lessen its environmental impact. Adopting sustainability not only tackles urgent ecological issues but also meets the rising consumer demand for ethically produced, high-quality textiles. Today's market is increasingly influenced by a heightened awareness of environmental and social responsibility, making sustainability crucial for maintaining brand relevance and consumer trust. Nonetheless, the transition to sustainability in luxury textiles is a challenging journey that requires investment in research, innovation, and the adoption of new technologies. Issues such as sourcing sustainable raw materials, implementing efficient production methods, and achieving circularity in design need to be tackled. Despite these challenges, the rewards are substantial, paving the way for a future where luxury and responsibility go hand in hand, ensuring the fashion industry's longevity while positively impacting the health of our planet.

References

1. Kumar, M. (2023). *Luxury fabrics: Meaning and definition*. https://www.tutorialspoint.com/luxury-fabrics-meaning-and-definition
2. Dull, C. (2023). *17 Luxury fabrics you need to know*. The Modern Directory. https://themoderndirectory.com/industry-insights/17-luxury-fabrics-you-need-to-know/
3. Liverpool. (2024). *Exclusive and rare fabrics in the world*. Signature Bespoke. https://signaturebespoke.co.uk/exclusive-and-rare-fabrics-in-the-world/#:~:text=Derived%20from%20Vicu%C3%B1as%2C%20small%2C%20similar,of%20wool%20in%20the%20world
4. Andruid. (2024). *30+ Luxury fabrics writers & world-builders should know*. Writing Games. https://writing-games.com/luxury-fabrics-for-crafting-wealthy-characters-a-guide-for-writers-and-world-builders/#:~:text=some%20quick%20definitions.-,What%20are%20luxury%20fabrics%3F,is%20proportional%20as%20income%20rises
5. Pauline, A., & Stenback, R. (2021). *Luxury fabrics: Worth every penny*. Stitch Fix Style. https://www.stitchfix.com/women/blog/fashion-tips/luxury-high-fashion-fabrics/#:~:text=What%20are%20Luxury%20Fabrics%3F,with%20a%20higher%20price%20tag
6. Tissura. (n.d.). *Luxury fabric: Fashion and designer sequined silk, lace and tulle fabric*. https://tissura.com/articles/luxury-fabrics
7. Kaur, S. (2023). *The rich tapestry of luxury fabrics: A historical evolution*. Sahni Fabs. https://www.sahnifabrics.com/blogs/news/the-rich-tapestry-of-luxury-fabrics-a-historical-evolution
8. Zoumpoulias, C. (2023). *Towards a sustainable fashion and luxury sector*. Deloitte United Kingdom. https://www.deloitte.com/uk/en/Industries/consumer/blogs/2023/towards-a-sustainable-fashion-and-luxury-sector.html
9. Bhandari, N., Garza-Reyes, J. A., Rocha-Lona, L., Kumar, A., Naz, F., & Joshi, R. (2022). Barriers to sustainable sourcing in the apparel and fashion luxury industry. *Sustainable Production and Consumption, 31*, 220–235. https://doi.org/10.1016/j.spc.2022.02.007
10. Ellis, K. (2024). *Report on the luxury textile market and its growth prospects*. Luxonomy. https://luxonomy.net/report-on-the-luxury-textile-market-and-its-growth-prospects/
11. Abbate, S., Centobelli, P., Cerchione, R., Nadeem, S. P., & Riccio, E. (2024). Sustainability trends and gaps in the textile, apparel and fashion industries. *Environment, Development and Sustainability, 26*(2), 2837–2864. https://doi.org/10.1007/s10668-022-02887-2
12. Todeschini, B. V., Cortimiglia, M. N., Callegaro-de-Menezes, D., & Ghezzi, A. (2017). Innovative and sustainable business models in the fashion industry: Entrepreneurial drivers,

opportunities, and challenges. *Business Horizons, 60*(6), 759–770. https://doi.org/10.1016/j.bushor.2017.07.003

13. Wu, J. X., & Li, L. (2019). Sustainability initiatives in the fashion industry. In *Fashion industry-An itinerary between feelings and technology*. IntechOpen. https://doi.org/10.5772/intechopen.87062
14. Bansal, A., & Attimukat. (2024). *Green chemistry: Innovations for sustainable manufacturing and resource management.* Inti Media. https://doi.org/10.9734/bpi/mono/978-81-973809-6-9/ch1
15. Tegegne, W., Haile, A., Zeleke, Y., Temesgen, Y., Bantie, H., & Biyable, S. (2024). Natural dyeing and anti-bacterial finishing of cotton fabric with extracts from Justicia schimperiana leaf extract: A step towards sustainable dyeing and finishing. *International Journal of Sustainable Engineering, 17*(1), 1–10. https://doi.org/10.1080/19397038.2023.2301702
16. De Smet, D., Weydts, D., & Vanneste, M. (2015). Environmentally friendly fabric finishes. In *Sustainable apparel* (pp. 3–33). Woodhead Publishing. https://doi.org/10.1016/B978-1-78242-339-3.00001-7
17. Billah, S. M. R. (2018). Textile coatings. In M. Jafar Mazumder, H. Sheardown, & A. Al-Ahmed (Eds.), *Functional Polymers* (Polymers and polymeric composites: A reference series). Springer. https://doi.org/10.1007/978-3-319-92067-2_30-1
18. Karthik, T., Rathinamoorthy, R., & Ganesan, P. (2015). Sustainable luxury natural fibers: Production, properties, and prospects. In *Environmental footprints and eco-design of products and processes* (pp. 59–98). Springer. https://doi.org/10.1007/978-981-287-633-1_4
19. Bilen, U. (2019). The effect of linen and linen blends on the comfort properties of bedding fabrics. *Journal of Natural Fibers, 18*(3), 430–441. https://doi.org/10.1080/15440478.2019.1624997
20. Santosh, G., Pavan, M., Praveenkumar, A., Om, P., Raigar P., Halladakeri Rakshith S. R., Gowda Pradeep K., Gurjeet S., Meenakshi T., Pusarla, S., Amandeep, Dinesh Kumar S. (2024). Fashion meets science: how advanced breeding approaches could revolutionize the textile industry. Critical Reviews in Biotechnology, 44(8), 1653–1679. https://doi.org/10.1080/07388551.2024.2314309
21. Crini, G., Lichtfouse, E., Chanet, G., & Morin-Crini, N. (2020). Traditional and new applications of hemp. In *Sustainable agriculture reviews* (pp. 37–87). Springer. https://doi.org/10.1007/978-3-030-41384-2_2
22. Martinko, K. (2020, September 17). *What is piñatex, and could it replace animal leather?* Treehugger. https://www.treehugger.com/what-is-pinatex-5074666
23. Camilli, S. (2023, May 16). *What is mushroom leather & why do people love it?* Immaculate Vegan. https://immaculatevegan.com/blogs/magazine/why-is-everyone-mad-about-mushroom-leather?srsltid=AfmBOooHQYuTXN1VzjgkKX3FOKWaDZJgKngEaDUHOUgIDwV71VLbeu5J
24. Shah, D. (2023). *Sustainable fibres for a greener fashion industry.* Fibre2Fashion. https://www.fibre2fashion.com/industry-article/9636/sustainable-fibres-for-a-greener-fashion-industry
25. Kiron, M. I. (2012). *Milk or casein fiber: Properties, manufacturing process, benefits and uses*. Textile Learner. https://textilelearner.net/milk-or-casein-fiber-properties/
26. *Regenerated fibres—Textile manufacturing and testing*. (n.d.). https://ebooks.inflibnet.ac.in/hsp08/chapter/regenerated-fibres/
27. Wendler, F., Schulze, T., Bauer, J., & Redlingshöfer, B. (2024). Bioactive lyocell fibers with inherent antibacterial, antiviral and antifungal properties. *Molecules, 29*(9), 2054. https://doi.org/10.3390/molecules29092054
28. CG Habitats. (n.d.). *RepReve®: Recycled board shorts.* https://cghabitats.com/pages/repreve-turning-plastic-bottles-into-board-shorts?ref=techpacker.com
29. Ali, M. A., & Sarwar, M. I. (2010). *Sustainable and environmental friendly fibers in textile fashion: A study of organic cotton and bamboo fibers* (University of Borås; Master Thesis 2010.9.14, University of Borås). https://www.diva-portal.org/smash/get/diva2:1312071/FULLTEXT01.pdf

30. Blackburn, R. S. (2009). *Sustainable textiles: Life cycle and environmental impact* (1st ed.). Woodhead Publishing.
31. Shahzad, A. (2011). Hemp fiber and its composites—A review. *Journal of Composite Materials, 46*(8), 973–986. https://doi.org/10.1177/0021998311413623
32. Sewport. (2019, March 6). *What is bamboo fabric: Properties, how its made and where.* https://sewport.com/fabrics-directory/bamboo-fabric
33. Imadi, S. R., Mahmood, I., & Kazi, A. G. (2014). Bamboo fiber processing, properties, and applications. In *Biomass and bioenergy* (pp. 27–46). Springer. https://doi.org/10.1007/978-3-319-07641-6_2
34. Shirke A. (2025). Bamboo Fabrics: Sustainable, Eco-Friendly textile. https://www.abhijeet-shirke.in/bamboo-fabrics/
35. Textilecoach. (2021). *Banana fiber: Textile fibers.* https://www.textilecoach.net/post/banana-fiber
36. Karthik, T., & Rathinamoorthy, R. (2017). Sustainable silk production. In *Sustainable fibres and textiles* (pp. 135–170). Woodhead Publishing. https://doi.org/10.1016/B978-0-08-102041-8.00006-8
37. Gull, A., Ganaie, M. A., & Chanotra, S. (2022). Trends in organic moriculture: An approach for revitalizing silk production. *Inventum Biologicum: An International Journal of Biological Research, 2*(2), 50–54. https://journals.worldbiologica.com/ib/article/view/15
38. Huang, W., Ling, S., Li, C., Omenetto, F. G., & Kaplan, D. L. (2018). Silkworm silk-based materials and devices generated using bio-nanotechnology. *Chemical Society Reviews, 47*(17), 6486–6504.
39. David, G. (2024). *Spider silk is woven into the future of sustainability.* SynBioBeta. https://www.synbiobeta.com/read/spider-silk-is-woven-into-the-future-of-sustainability
40. AMSilk. (n.d.). https://www.amsilk.com/
41. Spintex. (2020). https://www.spintex.co.uk/
42. Lee, K. E., & Sanders, E. A. (2015). Hanji, the mulberry paper yarn, rejuvenates nature and the sustainable fashion industry of Korea. In *Environmental footprints and eco-design of products and processes* (pp. 159–184). https://doi.org/10.1007/978-981-10-0111-6_7
43. Ananas Anam. (2022, November 8). Ananas Anam. https://www.ananas-anam.com/about-us/
44. Watson & Wolfe. (2024, April 1). *What is mushroom leather?* https://www.watsonwolfe.com/2020/02/08/what-is-mushroom-leather/
45. Varkki, A. (2024). *Mushroom leather.* Von Baer. https://vonbaer.com/blogs/blog/mushroom-leather?srsltid=AfmBOoqD3NfgtWRDFkiDrVDXyxS-Ht0xkwAUpNTPRP-PodPn5Ysl77bj
46. Ata. (2020). *Microalgae used to develop materials for fabric.* Textile Technology Source. https://textiletechsource.com/2020/05/05/microalgae-used-to-develop-materials-for-fabric/
47. Purton, N. (2023). *3 material innovators using algae to paint greener picture for fashion.* Sustainable Brands. https://sustainablebrands.com/read/materials-packaging/3-material-innovators-algae-greener-fashion
48. Cairns, R. (2021, October 26). *This vegan 'tech-style' startup wants to make clothes using algae.* CNN Style. https://edition.cnn.com/style/article/israel-algae-environment-textiles-spc-intl-hnk/index.html
49. Milkteds. (2020). *Milk fibre.* https://milkteds.africa/about/milk-fibre/
50. Srivastava, S. (2023). *Hey, what's milk fabric.* https://www.nonameglobal.com/post/milk-fabric-the-eco-friendly-fashion-revolution
51. Focus, T. (2024, March 28). *Fiber from Milk Byproducts—a New Dimension.* Textile Focus. https://textilefocus.com/fiber-from-milk-byproducts-a-new-dimension/
52. Baloyi, R. B., Gbadeyan, O. J., Sithole, B., & Chunilall, V. (2023). Recent advances in recycling technologies for waste textile fabrics: A review. *Textile Research Journal, 94*(3–4), 508–529. https://doi.org/10.1177/00405175231210239
53. Pavan M. and Samant L. (2022). Regenerated milk fiber: An approach towards green textiles. Just Agriculture e-Newsletter, 3(4), 6.

54. Lenzing. (2024). *Lenzing unveils the new potential of TENCEL™ Lyocell fibers as alternative material for stretch fabrics*. Lenzing—Innovative by Nature. https://www.lenzing.com/newsroom/press-releases/press-release/lenzing-unveils-the-new-potential-of-tenceltm-lyocell-fibers-as-alternative-material-for-stretch-fabrics
55. Daks India Industries. (2022). *Lyocell: The Sustainable Fabric*. Daks India Industries. https://daksindia.com/lyocell-fabric
56. CottonWorks™. (2024, April 9). *Recycled cotton*. https://cottonworks.com/en/topics/sustainability/cotton-sustainability/recycled-cotton/
57. Textil Santanderina. (2023, March 2). *Recycled cotton*. https://textilsantanderina.com/textile-solutions/recycled-cotton/
58. SustainYourStyle. (n.d.). *Recycled polyester*. https://www.sustainyourstyle.org/en/recycled-polyester2
59. Weavabel. (2024, March 27). *What are the pros and cons of recycled polyester?* https://www.weavabel.com/blog/what-are-the-pros-and-cons-of-recycled-polyester#:~:text=Recycled%20polyester%20is%20just%20as,energy%20compared%20to%20virgin%20polyester
60. *Recycled nylon: What it is, how it's made, and why we use it*. (n.d.). https://ylxgear.com/blog/recycled-nylon-what-how-why?srsltid=AfmBOoo2Q08qs_CRDV_SXiYpCVkRJZHjaRu2R8HH4MQRPAsyxI2Vpus6
61. Cariki. (n.d.). *Is recycled nylon sustainable? 15 eco pros and cons*. Cariki. https://cariki.co.uk/blogs/the-green-road/recycled-nylon-sustainable#:~:text=Yes%2C%20recycled%20nylon%20is%20considered,polyester%20or%20even%20regular%20cotton
62. Islam Kiron, M. (2022). *New fibers in the textile industry*. Textile Learner. https://textilelearner.net/new-fibers-in-the-textile-industry/
63. Davis, R. (2023, January 30). *Rachael Davis*. https://www.textileworld.com/textile-world/features/2023/01/the-future-of-fibers/
64. Omnexus. (2021, February 19). *Polyethylene furanoate (PEF) as a renewable bioplastic*. https://omnexus.specialchem.com/tech-library/article/polyethylene-furanoate-pef-bioplastic#:~:text=There%20are%20several%20applications%20where,and%20recyclable%20alternative%20to%20PET
65. Arputharaj, A., Raja, A. S. M., & Saxena, S. (2016). Developments in sustainable chemical processing of textiles. *Green Fashion, 1*, 217–252. https://doi.org/10.1007/978-981-10-0111-6_9
66. Periyasamy, A. P., Venkataraman, M., Kremenakova, D., Militky, J., & Zhou, Y. (2020). Progress in sol-gel technology for the coatings of fabrics. *Materials, 13*(8), 1838. https://doi.org/10.3390/ma13081838
67. Samanta, K. K., Basak, S., & Chattopadhyay, S. K. (2015). Specialty chemical finishes for sustainable luxurious textiles. *Handbook of Sustainable Luxury Textiles and Fashion, 1*, 145–184. https://doi.org/10.1007/978-981-287-633-1_7
68. wax are commonly found in natural M., Samant, L., Mahajan, S., & Kaur, M. (2024). Role of chemicals in textile processing and its alternatives. In *Climate action through eco-friendly textiles* (pp. 55–72). Springer Nature. https://doi.org/10.1007/978-981-99-9856-2_5
69. Ammayappan, L., Jose, S., & Raj, A. A. (2015). Sustainable production processes in textile dyeing. In *Environmental footprints and eco-design of products and processes* (pp. 185–216). Springer. https://doi.org/10.1007/978-981-10-0111-6_8
70. Mojsov, K. (2011). Application of enzymes in the textile industry: A review. In *II International Congress "Engineering, Ecology and Materials in the Processing Industry", Proceedings, 2011, 09–11 March, Jahorina, Bosnia and Hercegovina* (pp. 230–239) https://eprints.ugd.edu.mk/id/eprint/1381
71. Pavan M, Mahajan, S., Lakshmanan, A., Nageshkumar, T., Midha, V., Manjunatha, B. S., & Grewal, S. (2024). Optimization of an alkali modification protocol on Crotalaria juncea fibre and its characterization for technical textile applications. Industrial Crops and Products, 211, 118275. https://doi.org/10.1016/j.indcrop.2024.118275
72. Couto, S. R., & Toca-Herrera, J. L. (2006). Lacasses in the textile industry. *Biotechnology and Molecular Biology Reviews, 1*(4), 115–120.

73. Affat, S. S. (2021). Classifications, advantages, disadvantages, toxicity effects of natural and synthetic dyes: A review. *University of Thi-Qar Journal of Science, 8*(1), 130–135.
74. Pizzicato, B., Pacifico, S., Cayuela, D., Mijas, G., & Riba-Moliner, M. (2023). Advancements in sustainable natural dyes for textile applications: A review. *Molecules, 28*(16), 5954. https://doi.org/10.3390/molecules28165954
75. Saxena, S., & Raja, A. S. M. (2014). Natural dyes: Sources, chemistry, application and sustainability issues. In *Roadmap to sustainable textiles and clothing: Eco-friendly raw materials, technologies, and processing methods* (pp. 37–80). Springer. https://doi.org/10.1007/978-981-287-065-0_2
76. Adeel, S., Rafi, S., Mustaan, M. A., Salman, M., & Ghaffar, A. (2018). Animal based natural dyes: A short review. In *Handbook of renewable materials for coloration and finishing* (pp. 41–74). Wiley. https://doi.org/10.1002/9781119407850.ch4
77. Yadav, S., Tiwari, K. S., Gupta, C., Tiwari, M. K., Khan, A., & Sonkar, S. P. (2023). A brief review on natural dyes, pigments: Recent advances and future perspectives. *Results in Chemistry, 5*, 100733. https://doi.org/10.1016/j.rechem.2022.100733
78. Yusuf, M., Shahid, M., Khan, S. A., Khan, M. I., Islam, S. U., Mohammad, F., & Khan, M. A. (2013). Eco-dyeing of wool using aqueous extract of the roots of Indian madder (Rubia cordifolia) as natural dye. *Journal of Natural Fibers, 10*(1), 14–28. https://doi.org/10.1080/15440478.2012.738026
79. Shepherd Textiles. (2024). *Sappanwood extract*. https://shepherdtextiles.com/guides/sappanwood-extract#:~:text=The%20heartwood%20of%20the%20sappanwood,country%20of%20Brazil%20its%20name
80. Deveoglu, O., Karadag, R., Torgan, E., & Yildiz, Y. (2020). Examination of dyeing properties of the dyed cotton fabrics with barberry (Berberis vulgaris L.). *Journal of Natural Fibers, 17*(8), 1089–1098. https://doi.org/10.1080/15440478.2018.1558143
81. Bolognesi, V. J., & Garcia, C. E. (2018). Annatto carotenoids as additives replacers in meat products. In *Alternative and replacement foods* (pp. 355–384). Academic Press. https://doi.org/10.1016/B978-0-12-811446-9.00012-5
82. Hübert, T., Tiebe, C., & Banach, U. (2016). Electronic noses for the quality control of spices. In *Electronic noses and tongues in food science* (pp. 115–124). Academic Press. https://doi.org/10.1016/B978-0-12-800243-8.00012-3
83. Hasan, K. F., Hridam, D., Rahman, M. M., Morshed, M. N., Al Azad, S., & Genyang, C. (2016). A review on antibacterial coloration agent's activity, implementation & efficiency to ensure the ecofriendly & green textiles. *American Journal of Polymer Science AND Engineering, 4*(1), 39–59.
84. Dhingra, S. (2016). Dyeing with Morinda citrifolia: In pursuit of sustainable future. *Textile Society of America Symposium Proceedings, 957*, 75–86. http://digitalcommons.unl.edu/tsaconf/957
85. Abdeldaiem, M. H. (2013). Use of yellow pigment extracted from turmeric (Curcuma longa) rhizomes powder as natural food preservative and colorant. *Food Science and Quality Management, 22*, 56–68.
86. Rahman, N. A., Tajuddin, R., & Tumin, S. M. (2013). Optimization of natural dyeing using ultrasonic method and biomordant. *International Journal of Chemical Engineering and Applications, 4*(3), 161.
87. Rovira, J., Nadal, M., Schuhmacher, M., & Domingo, J. L. (2015). Human exposure to trace elements through the skin by direct contact with clothing: Risk assessment. *Environmental Research, 140*, 308–316. https://doi.org/10.1016/j.envres.2015.03.032
88. Boutrup, J., & Ellis, C. (2018). *The art and science of natural dyes: Principles, experiments, and results*. Schiffer Publishing.
89. Benli, H. (2024). Bio-mordants: A review. *Environmental Science and Pollution Research, 31*(14), 20714–20771. https://doi.org/10.1007/s11356-024-32174-8
90. Verma, H., Sharma, A. K., Rani, A., Thiyam, G., & Pavan, M. (2023). Optimization of the process parameters for dyeing silk fabric with a fungal pigment using RSM. *Sustainable Chemistry and Pharmacy, 35*, 101192. https://doi.org/10.1016/j.scp.2023.101192

91. Simsek Geyik, M., Efe, D., & Gormez, A. (2024). Identification of bacteria producing red pigments and their application in the textile industry. *Arabian Journal for Science and Engineering*. https://doi.org/10.1007/s13369-024-09290-1
92. Rather, L. J., Mir, S. S., Ganie, S. A., & Li, Q. (2023). Research progress, challenges, and perspectives in microbial pigment production for industrial applications—A review. *Dyes and Pigments, 210*, 110989. https://doi.org/10.1016/j.dyepig.2022.110989
93. Mapari, S. A., Thrane, U., & Meyer, A. S. (2010). Fungal polyketide azaphilone pigments as future natural food colorants? *Trends in Biotechnology, 28*(6), 300–307. https://doi.org/10.1016/j.tibtech.2010.03.004
94. Mishra, A. (2007). *Natural dyes from pseudomonas fluorescens: Extraction, characterization and optimization of dyeing process for textiles* (PhD thesis). GB Pant University of Agriculture and Technology, Pantnagar
95. Basant, T. (2012). *Natural dye from monascus purpureus: Optimization of dyeing process and evaluation of its colour fastness and functional properties for cotton and silk fabric* (PhD thesis). G B Pant University of Agriculture and Technology, Pantnagar
96. Bhargava, D. (2012). *Microbial dyes from pseudomonas fluorescens and pseudomonas aeruginosa: Extraction and optimization of dyeing process for textiles* (PhD thesis). G B Pant University of Agriculture and Technology, Pantnagar
97. Thakker, A. M., & Sun, D. (2023). Inks for digital printing of textiles. *Journal of the Textile Institute, 115*, 2377–2390. https://doi.org/10.1080/00405000.2023.2289975
98. Lee, W. K., Sung, E. J., Moon, J. R., Ahn, I. Y., Yoon, K. H., Park, Y. C., & Kim, J. H. (2021). Performance analysis of ink for digital textile printing using natural indigo. *Textile Coloration and Finishing, 33*(4), 202–209. https://doi.org/10.5764/TCF.2021.33.4.202
99. Zahid, M., Heredia-Guerrero, J. A., Athanassiou, A., & Bayer, I. S. (2017). Robust water repellent treatment for woven cotton fabrics with eco-friendly polymers. *Chemical Engineering Journal, 319*, 321–332. https://doi.org/10.1016/j.cej.2017.03.006
100. Mohamed, A. L., Sedik, A., Mosaad, M. M., & Othman, H. A. (2023). Natural Oils\Silicon based materials for imparting new properties to cellulose based fabrics. *Results in Chemistry, 5*, 100812. https://doi.org/10.1016/j.rechem.2023.100812
101. Bashari, A., Salehi, K. A. H., & Salamatipour, N. (2020). Bioinspired and green water repellent finishing of textiles using carnauba wax and layer-by-layer technique. *The Journal of the Textile Institute, 111*(8), 1148–1158. https://doi.org/10.1080/00405000.2019.1686881
102. Sharif, R., Mohsin, M., Qutab, H. G., Saleem, F., Bano, S., Nasir, R., & Wahlah, A. (2023). Durable water and oil repellents along with green chemistries: An overview. *Chemical Papers, 77*(7), 3547–3560. https://doi.org/10.1007/s11696-023-02763-x
103. Mao, N., & Du, M. (2018). Sol–gel-based treatments of textiles for water repellence. In *Waterproof and water repellent textiles and clothing* (pp. 233–265). Woodhead Publishing. https://doi.org/10.1016/B978-0-08-101212-3.00009-5
104. Zhao, Q., Wu, L. Y. L., Huang, H., & Liu, Y. (2016). Ambient-curable superhydrophobic fabric coating prepared by water-based non-fluorinated formulation. *Materials & Design, 92*, 541–545. https://doi.org/10.1016/j.matdes.2015.12.054
105. Onar, N., & Mete, G. (2016). Development of water repellent cotton fabric with application of ZnO, Al_2O_3, TiO_2 and ZrO_2 nanoparticles modified with ormosils. *Textile and Apparel, 26*(3), 295–302.
106. Edward, N. W., & Goswami, P. (2018). Plasma-based treatments of textiles for water repellency. In *Waterproof and water repellent textiles and clothing* (pp. 215–232). Woodhead Publishing. https://doi.org/10.1016/B978-0-08-101212-3.00008-3
107. Sheraz, M., Choi, B., & Kim, J. (2023). Enhancing textile water repellency with octadecyltrichlorosilane (OTS) and hollow silica nanoparticles. *Polymers, 15*(20), 4065. https://doi.org/10.3390/polym15204065
108. Dashairya, L., Barik, D. D., & Saha, P. (2019). Methyltrichlorosilane functionalized silica nanoparticles-treated superhydrophobic cotton for oil–water separation. *Journal of Coatings Technology and Research, 16*, 1021–1032. https://doi.org/10.1007/s11998-018-00177-z

109. Dehabadi, V. A., Buschmann, H. J., & Gutmann, J. S. (2013). Durable press finishing of cotton fabrics: An overview. *Textile Research Journal, 83*(18), 1974–1995. https://doi.org/10.1177/0040517513483857
110. Mensah, R. A., Shanmugam, V., Narayanan, S., Renner, J. S., Babu, K., Neisiany, R. E., et al. (2022). A review of sustainable and environment-friendly flame retardants used in plastics. *Polymer Testing, 108*, 107511.
111. Sonnier, R., Taguet, A., Ferry, L., Lopez-Cuesta, J. M., Sonnier, R., Taguet, A., et al. (2018). Towards bio-based flame retardant polymers. In *Biobased flame retardants* (pp. 33–72). Springer International Publishing. https://doi.org/10.1007/978-3-319-67083-6
112. Oualha, M. A., Omri, N., Oualha, R., Nouioui, M. A., Abderrabba, M., Amdouni, N., & Laoutid, F. (2019). Development of metal hydroxide nanoparticles from eggshell waste and seawater and their application as flame retardants for ethylene-vinyl acetate copolymer (EVA). *International Journal of Biological Macromolecules, 128*, 994–1001. https://doi.org/10.1016/j.ijbiomac.2019.02.065
113. Krishnadevi, K., & Selvaraj, V. (2016). Biowaste material reinforced cyanate ester based epoxy composites for flame retardant applications. *High Performance Polymers, 28*(8), 881–894. https://doi.org/10.1177/0954008315605554
114. Shah, A. U. R., Prabhakar, M. N., Saleem, M., & Song, J. I. (2017). Development of bio-waste encapsulated polypropylene composites: Thermal, optical, dielectric, flame retardant, mechanical, and morphological properties. *Polymer Composites, 38*(2), 236–243. https://doi.org/10.1002/pc.23580
115. Ménard, R., Negrell, C., Ferry, L., Sonnier, R., & David, G. (2015). Synthesis of biobased phosphorus-containing flame retardants for epoxy thermosets comparison of additive and reactive approaches. *Polymer Degradation and Stability, 120*, 300–312. https://doi.org/10.1016/j.polymdegradstab.2015.07.015
116. Ménard, R., Negrell-Guirao, C., Ferry, L., Sonnier, R., & David, G. (2014). Synthesis of bio-based phosphate flame retardants: Characterization of flame retardancy on epoxy thermosets. *Pure and Applied Chemistry, 86*(11), 1637–1650. https://doi.org/10.1515/pac-2014-0703
117. Mauldin, T. C., Zammarano, M., Gilman, J. W., Shields, J. R., & Boday, D. J. (2014). Synthesis and characterization of isosorbide-based polyphosphonates as biobased flame-retardants. *Polymer Chemistry, 5*(17), 5139–5146. https://doi.org/10.1039/C4PY00591K
118. Huang, Y., Ma, T., Wang, Q., & Guo, C. (2019). Synthesis of biobased flame-retardant carboxylic acid curing agent and application in wood surface coating. *ACS Sustainable Chemistry & Engineering, 7*(17), 14727–14738. https://doi.org/10.1021/acssuschemeng.9b02645
119. Zhu, Z. M., Shang, K., Wang, L. X., & Wang, J. S. (2019). Synthesis of an effective bio-based flame-retardant curing agent and its application in epoxy resin: Curing behavior, thermal stability and flame retardancy. *Polymer Degradation and Stability, 167*, 179–188. https://doi.org/10.1016/j.polymdegradstab.2019.07.005
120. Ye, G., Huo, S., Wang, C., Shi, Q., Liu, Z., & Wang, H. (2021). One-step and green synthesis of a bio-based high-efficiency flame retardant for poly (lactic acid). *Polymer Degradation and Stability, 192*, 109696.
121. Kiron, M I. (2013). *Mothproofing finishing process of fabric*. Textile Learner. https://textilelearner.net/moth-proofing-finish-of-fabric/
122. MothPrevention. (2022, February 17). *What is the best natural moth repellent?* https://www.moth-prevention.com/blogs/the-art-of-prevention/natural-moth-repellent
123. Medha, K., Rajna, S., Devi, L. J., Samant, L., & Jose, S. (2021). A comprehensive review on moth repellent finishing of woolen textiles. *Journal of Cultural Heritage, 49*, 260–271. https://doi.org/10.1016/j.culher.2021.02.006
124. Sro, C. T. (n.d.). *Special textiles*. https://www.vubas.cz/en/special-textiles
125. Kheoni. (n.d.). *Therapeutic eye pillow*. https://www.kheoni.com/products/therapeutic-eye-pillow

126. Khedkar, S. S., & Mallick, A. S. C. (2020). Aroma textiles. In *Advances in functional and protective textiles* (pp. 169–190). Woodhead Publishing. https://doi.org/10.1016/B978-0-12-820257-9.00008-4
127. Singh, N., Yadav, M., Khanna, S., & Sahu, O. (2017). Sustainable fragrance cum antimicrobial finishing on cotton: Indigenous essential oil. *Sustainable Chemistry and Pharmacy, 5*, 22–29. https://doi.org/10.1016/j.scp.2017.01.003
128. Cavaco-Paulo, A., & Gubitz, G. (Eds.). (2003). *Textile processing with enzymes* (Vol. 29). Woodhead Publishing.
129. Roy Choudhury, A. K. (2014). Sustainable textile wet processing: Applications of enzymes. In *Roadmap to sustainable textiles and clothing: Eco-friendly raw materials, technologies, and processing methods* (pp. 203–238). Springer. https://doi.org/10.1007/978-981-287-065-0_7
130. Riva, S. (2006). Laccases: Blue enzymes for green chemistry. *Trends in Biotechnology, 24*(5), 219–226. https://doi.org/10.1016/j.tibtech.2006.03.006
131. Mogilireddy, V. (2018). *Sustainable dyeing innovations: Greener ways to color textiles*. https://www.prescouter.com/2018/11/sustainable-dyeing-innovations-greener-ways-color-textiles/
132. Recycrom™. (n.d.). *Officina39*. https://recycrom.com/
133. Huntsman. (n.d.). *Sustainable products*. https://www.fibre2fashion.com/sustainability-plus/huntsman/products.asp
134. Gabardi, C. (2024). *Fashion and sustainability: How luxury brands are innovating with eco-friendly materials*. Luxonomy. https://luxonomy.net/fashion-and-sustainability-how-luxury-brands-are-innovating-with-eco-friendly-materials/#:~:text=Gucci%20has%20also%20committed%20to,cotton%20in%20its%20latest%20collections
135. Deeley, R. (2023). *Case study: Fashion's race for new materials*. The Business of Fashion. https://www.businessoffashion.com/case-studies/sustainability/materials-innovation-textiles-recycling-production/
136. Lodigiani, I. (2023). *Prada keeps on working with nylon—But searching for solutions*. The Lamp. https://lampoonmagazine.com/article/2023/02/23/prada-keeps-on-working-with-nylon-but-searching-for-solutions/
137. Adidas. (2022). *How we turn plastic bottles into shoes: Our partnership with parley for the oceans*. https://www.adidas.co.in/blog/639412-how-we-turn-plastic-bottles-into-shoes-our-partnership-with-parley-for-the-oceans
138. Stella McCartney. (n.d.). *Fibres from forests*. https://www.stellamccartney.com/us/en/sustainability/fibres-from-forests.html?srsltid=AfmBOooHbUkcUuaXhX-TuONN51hik2vX_NBanEGwt0d5hvzS4fNXNj_3

Green Chemistry and Clean Technology in Luxury Textile Manufacturing

Manjulatha C., Surya Tejaswi Desu, Kshetrimayum Dolly Devi, Sirisha Deepthi Sornapudi, Kanchan Bisht, Surabhi Das, and Alka Goel

1 Introduction

The striking contrast between anthropogenic and natural chemical processes has driven a rapid escalation of ecological challenges. Although biochemical mechanisms have historically utilized readily available elements to create terrestrial matter, modern industries have extracted and dispersed elements globally in unprecedented ways. Lead, once isolated, now permeates the environment owing to industrial utilization. Synthetic compounds in pharmaceuticals, plastics, and pesticides are dissimilar from natural products and appear almost extraterrestrial [1, 2]. Industrial processes operate at significantly higher rates and on larger scales than natural biochemical processes, resulting in the rapid accumulation of elements and compounds in unexpected locations. The complexity and persistence of numerous synthetic compounds further challenge natural decomposition and recycling processes, exacerbating their environmental impacts [3].

Manjulatha C. (✉)
Acharya N. G. Ranga Agricultural University, Lam Guntur, Andhra Pradesh, India
e-mail: c.manjulatha@angrau.ac.in

S. T. Desu · K. Bisht · S. Das · A. Goel
G B Pant University of Agriculture and Technology, Pantnagar, Uttarakhand, India

K. D. Devi
Punjab Agricultural University, Ludhiana, Punjab, India

S. D. Sornapudi
Jayashankar Telangana Agricultural University, Hyderabad, Telangana, India

Sadhna et al. (eds.), *Crafting Sustainability in Luxury Textiles for a Zero-Waste Future*, Sustainable Textiles: Production, Processing, Manufacturing & Chemistry, https://doi.org/10.1007/978-3-031-86773-6_12

2 Impact of Fashion

The ecological consequences associated with the production and consumption of clothing are significant. Textile production requires significant coal [4], contributes to industrial water pollution, and utilizes a large proportion of global insecticides. Polyester fibre production is energy-intensive, with most emissions occurring during manufacturing [5]. However, consumer use also significantly contributes to clothing lifecycle emissions. Current clothing purchases have quadrupled compared with those two decades ago, with an estimated 80 billion garments produced annually [6–8]. In the UK, the average household's global footprint generates 1.5 tonnes of CO_2 emissions and a water footprint exceeding 200,000 L annually [9], which significantly contributes to waste generation.

Luxury firms have been subject to criticism because of their lack of sustainable practices and delayed responses to societal and environmental concerns [10]. The concept of luxury is categorized differently in the literature, considering aspects such as product/service provision, psychological/social ramifications, technical characteristics, and consumer perceptions, including exclusivity, product excellence, high pricing, individual emotional factors, and brand influence [11]. As the second most significant source of carbon emissions worldwide, the luxury fashion and textile industry is under pressure from governmental bodies, non-governmental organizations, consumers, and the media to reduce its environmental impact (Table 1) [13]. The diminishing value of luxury brands and increasing consumer preference for sustainable goods necessitate the incorporation of sustainable practices in the production process of luxury textile goods [14], with emerging technologies potentially facilitating sustainable transitions in luxury fashion [15].

3 Luxury Textiles and Sustainability

As introduced by the Brundtland Commission in 1987, the business sector embraced the notion of sustainable development. Confronted with numerous challenges, organizations now consider sustainable development essential [16]. Sustainability is characterized by the fulfilment of current human requirements while ensuring that future generations retain the capacity to satisfy their own needs [17]. In recent years, manufacturers of high-end products have begun to place greater emphasis on ecological and social considerations, alongside traditional factors such as branding, rarity, and product excellence [18]. The textile industry also recognizes the demand for green chemistry principles in addressing the environmental impact of conventional production methodologies. This paradigm shift aims to reduce harmful chemicals, minimize waste, and enhance the overall sustainability of textile products throughout their life cycle.

Table 1 Environmental impact caused by fibre [12]

Fibre	Renewable/non renewable	Biodegradability at product end of life	Ecological and social consequences	Utilization of water and resources	Energy usage
Cotton	Renewable	Biodegradable	Environmental pollution caused due to discharge of effluents from processing, usage of fertilizers and pesticides for crop protection. Finishing chemicals such as formaldehyde for permanent press finish. Low wages and forced labour, child labour	A significant volume of water, ranging from (8000–22,000 L/kg)	Less
Wool	Renewable	Biodegradable	Environmental pollution due to chemicals used for growth and maintenance of sheep. Toxic chemicals for wet processing. Methane gas pollution from sheep	Utilization of extensive agricultural land	Moderate to low
Silk	Renewable	Biodegradable	Environmental pollution caused due to disinfectants e.g., chlorine, lime, formaldehyde. Metallic salts for weighting of silk, effluents from dyes and other wet processing and finishing. Health hazards associated with the reeling industry and poor work hygiene. Child labour and low wages	The utilization of agricultural land for the cultivation of silk	Regulating temperature and humidity demands a considerable amount of energy. The drying of cocoons through the application of heated air or steam necessitates a significant amount of energy consumption
Polyester	Non renewable	Not	Pollution caused by non-biodegradable nature of polyester which taken several 100 years to disintegrate and the cause of plastic pollution microplastic pollution which enters into marine and land food chain system and causes solid waste GHG generation		

(continued)

Table 1 (continued)

Fibre	Renewable/non renewable	Biodegradability at product end of life	Ecological and social consequences	Utilization of water and resources	Energy usage
Nylon	Non renewable	Not	Environmental contamination arises from two main sources: • The production process emits nitrous oxide, a GHG that is more harmful than carbon CO_2 in terms of ozone depletion. Laundry of nylon garments produce microfiber pollution, impacting marine ecosystems	Uses monomers for polymerization of fibre	Energy intensive manufacturing process that can lead to global warming (three times energy usage compared to cotton)
Rayon	Renewable	Biodegradable	The primary environmental issues stem from: • Environment pollution resulting from the use of carbon disulfide and sodium hydroxide • The extraction of timber from ancient forests to obtain the necessary components for manufacturing viscose textiles	Extensive use of farm land which demands deforestation	Energy demanding Production processes
Acrylic	Non renewable	Not	The following factors contribute to environmental issues: • The emission of GHGs like CO_2 and SO_2 can trigger acid rain. Additionally, the production of non-hazardous waste materials includes discarded cans, paper products, plastics, and filter fabrics	Need nearly 10 kg of steam for the production of 1 kg fibre	Energy demanding Production processes

3.1 Green Chemistry Concept and Principles

Green chemistry focuses on innovative scientific and technological methods to reduce waste production. The core tenets of this discipline provide an upstream answer to various health, environmental, and economic issues associated with the industrial use of chemicals [19]. Cleaner production adopts a comprehensive strategy that integrates both the conceptual and applied components. The approach encompasses the entire lifecycle of products and processes, in order to reduce short and long-term hazards to environmental and human well-being. Green chemistry attempts to prevent environmental depletion at the ground level by emphasizing the role of chemical compounds, whereas clean technologies focus on the optimization of processes related to recycling separation, resource recovery, conservation, and the efficient utilization of raw materials, water, and energy. They also aim to improve production methods and manage or recycle waste that cannot be avoided. In this context, green chemistry emerges as a supportive strategy for clean technology, emphasizing principles derived from chemical engineering rather than solely from pure chemistry [20–22].

3.2 Principles of Green Chemistry

Twelve principles of green chemistry are (Fig. 1) [23, 24].

1. *Prevent waste*: It is more advantageous to avoid waste generation than to manage or discard it once it has been produced.
2. *Atom economy*: The development of synthetic procedures should aim to optimize the integration of all utilized materials into the end product.
3. *Less hazardous synthesis*: Whenever possible, chemical synthesis methods should be designed to create and use substances that are either harmless or minimally harmful to both human health and the environment.
4. *Design of benign chemicals*: The design of chemical products should prioritize maintaining functional effectiveness while minimizing their toxic properties.
5. *Benign solvents and auxiliaries*: The use of additional materials (such as solvents and separation agents) should be minimized whenever possible. If necessary, these substances should be nontoxic.
6. *Design for energy efficiency*: The environmental and economic implications of the energy demand should be acknowledged and reduced. Synthetic processes should be performed at standard temperature and atmospheric pressure.
7. *Use of renewable feedstocks*: Whenever feasible from a technical and economic standpoint, it is preferable to utilize renewable sources for raw materials or feedstocks rather than those that are finite.
8. *Reduce derivatives*: To minimize additional reagent usage and waste production, it is advisable to limit or eliminate superfluous chemical modifications.

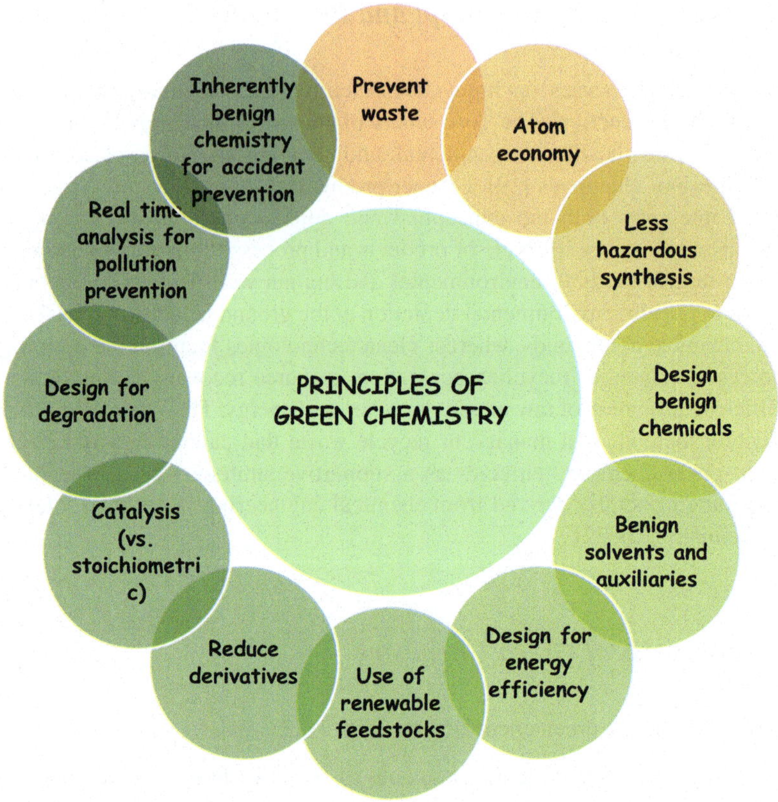

Fig. 1 Principles of green chemistry [23, 24]

This includes reducing the use of blocking groups, protection and deprotection processes, and temporary alterations to the physical or chemical properties.

9. *Catalysis (vs. stoichiometric)*: Reagents that catalyse reactions with high selectivity are preferable to those used in stoichiometric quantities.
10. *Design for degradation*: When creating chemical products, the primary focus should be on ensuring that they can decompose into harmless substances once their intended use is complete, thereby minimizing their enduring environmental impact.
11. *Real-time analysis for pollution prevention*: The ongoing development of improved analytical methods is essential for enabling continuous on-site supervision and control prior to the generation of dangerous substances.
12. *Inherently benign chemistry for accident prevention*: In chemical operations, the primary consideration in selecting chemicals and their forms should be the minimization of accident risk, including spills, explosions, and fires.

4 Raw Materials and Its Sustainability

Textile fibres are of paramount importance in the fashion and textile sectors. Natural fibres, including cotton, wool, flax, and jute, have been utilized in garment production for thousands of years, with a history spanning 5000 years [25]. Additional significant natural fibres include speciality animal hairs (llama, angora, alpaca, camel, and mohair), hemp, sisal, and silk [26]. The 1930s marked the emergence of synthetic fibres, commencing with the invention of rayon. Generally, textile fibres are categorized into two main groups: those that occur naturally and those that are manufactured synthetically [27]. While natural fibres are plant/animal/mineral-based, synthetic fibres are produced using petroleum (acrylic, nylon, etc.) or natural resources (like rayon and lyocell). Compared to synthetics, natural fibres such as flax, hemp, jute, wool, and nettle offer a wide range of benefits. They are renewable, biodegradable, non-abrasive, and naturally available [28, 29]. Furthermore, plant fibres, such as flax and hemp, are deemed more environmentally friendly because of their capacity to absorb carbon dioxide and release oxygen.

Cultivation of natural fibres like cotton is resource intensive process, that is, involves the use of water, land, and harmful chemicals. The practice has substantial ecological and societal consequences, encompassing contamination of soil, atmosphere, and water resources, as well as contributions to climate change through greenhouse gas (GHG) emissions and the ozone layer depletion.

Conversely, synthetic fibres require minimal water (approximately 1/100th) and no toxic chemicals. However, synthetic fibre production is energy-intensive; for instance, polymerization, spinning, and finishing polyester require 369–432 MJ/kg, whereas cotton production needs only 38–46 MJ/kg [30]. Unlike synthetic fibres produced from petroleum, which exhaust limited resources, natural fibres are sustainable and replenishable. Natural fibres also have an advantage in waste management because of their biodegradability. Being the most sustainable fibre, organic cotton has the least impact on environmental impact followed by flax. Polyester, polypropylene, and acrylic scored lower, making them less preferred in terms of their environmental impact (Table 2) [31].

4.1 *Sustainable Fibres*

The environmental depletion caused due to production and processing of textile fibres can be mitigated through the selection of sustainable alternatives (Table 3). This discussion explores several eco-friendly fibre options suitable for luxury textile manufacturing.

Table 2 Impact caused by various textile fibres on environment in their decreasing order [25]

	Energy usage	Water usage	Greenhouse gas emissions	Waste water effluents	Direct land use
Decreasing environmental impact ↓	Acrylic	Cotton	Wool	Wool	Wool
	Nylon Polyester Regenerated cellulose PLA/Cotton/ Lyoce ll Wool Natural Bast fibres (Nettle Hemp, Flax)	Silk Nylon Regenerated cellulosic Acrylic Hemp Wool Natural Bast based fibres Wool	Regenerated cellulosic Natural bast based fibres Nylon Polyester	Ramie Cotton Flax Hemp Viscose and modal Jute PLA Lyocell (Synthetic)	Ramie Cotton Flax Hemp Viscose and modal Jute PLA Lyocell (synthetic)

Table 3 Different Textile fibers (fibre groups) for environmental sustainability [12]

Different fibre groups	Fibres
Organic	Wild silk, Organic wool and cotton, Kala cotton
Other natural fibres	Hemp, mechanically extracted bamboo and banana, flax, Himalayan fibers (nettle, bhimal, kenaf and Rambans) and other leaf and bast fibres
Regenerated fibres	Bamboo, PLA (corn), lyocell, soy protein, casein, Seashell, chitin and chitosan
Other fibres	Lotus, mechanically extracted bamboo, banana, milk protein and soy protein fibres
Recycled natural fibres	Cotton, wool, hemp and flax
Recycled synthetic fibres	Polyester, nylon and polypropylene recycled mechanically or chemically
Recycled blends	Blends of natural and synthetic fibres

4.1.1 Organic Cotton and Naturally Coloured Cotton

Organic cotton is cultivated and certified in accordance with organic farming standards without the use of synthetic fertilizers, pesticides, herbicides, growth regulators, or defoliants. This methodology prohibits the use of genetically modified seeds and deleterious agrochemical pesticides and fertilizers. Organic cotton cultivation employs environmentally sustainable techniques and materials, enhances soil health, and mitigates the application of harmful pesticides and fertilizers (Fig. 2). Sustainably produced organic cotton is intended to conserve natural resources without chemical intervention. The primary distinction between organic and conventional cotton is the substantial environmental advantages offered by the former.

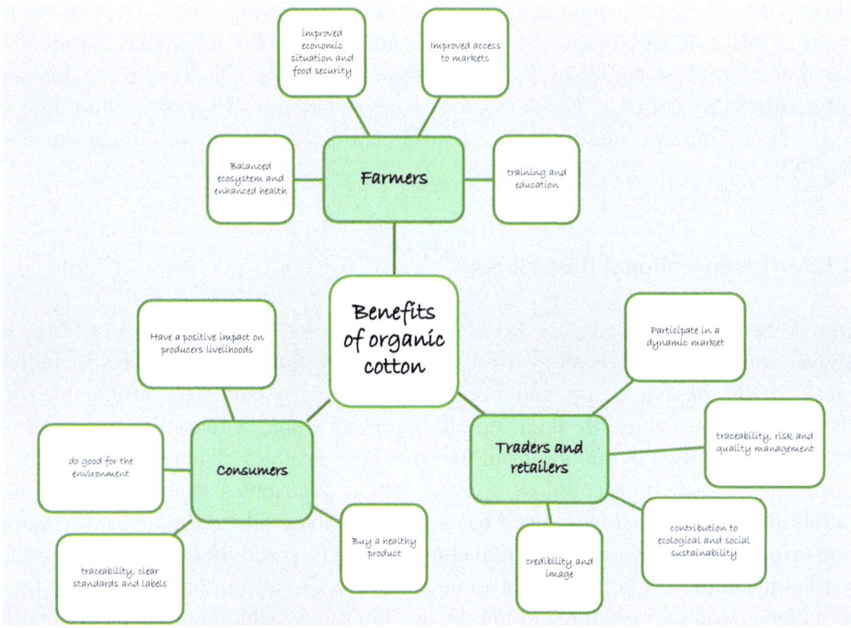

Fig. 2 Benefits of organic cotton [26]

Fig. 3 Naturally coloured cotton(left), Eri silk (right) [28, 29]

4.1.2 Naturally Coloured Cotton

Since antiquity, both coloured cotton (Fig. 3) and white-lined cotton have been extant, with the cultivation of coloured cotton dating back to 2500 B.C. However, advancements in dyeing techniques have resulted in a shift towards white cotton, leading to the abandonment of naturally coloured varieties. In recent years, coloured cotton has garnered renewed interest because of its environmentally advantageous characteristics. The German ban on azo dyes highlights their harmful impacts on both human well-being and ecosystems. These substances are known to trigger various health issues, including allergic reactions, itching, and cancer development, while causing significant environmental harm. This heightened awareness has precipitated a resurgence of organic cotton cultivation. The increasing demand for eco-friendly cotton is most effectively met by organically grown coloured cotton, which eliminates the need for harmful chemicals during dyeing and processing [27].

4.1.3 Unconventional Bast Fibres

Researchers have explored the possibility of utilizing underutilized bast fibres in conventional textiles as well as for technical applications. Natural fibres including ramie, nettle, *bhimal*, hemp, and flax have significant potential for sustainable textile production owing to their unique properties and environmental benefits. Production involves extraction from the outer cell layers of plant stems and is characterized by their high cellulose content, which contributes to their strength and durability. The increasing demand for sustainable materials in the fields of fashion and textiles, driven by environmental concerns and the need for biodegradable alternatives to synthetic fibres, has motivated researchers to invent the process of extraction conversion of bast fibres to textiles for high-end fashion and technical textile applications [30].

4.2 Sustainable Silk

Silk, an irreplaceable luxury fibre, is a key raw material for various luxury products and is consistently associated with wealth. Silk producers eliminate worms by submerging cocoons in scalding water or placing them in heated ovens before the moths emerge, thus preventing damage caused by moth excretion. The killing of the moth makes the silk fibre unsustainable [31]. Some of the alternative methods of extracting silk fibres without killing moths are mentioned here.

4.2.1 Organic Silk

The production of organic silk is both eco-friendly and ethical. It yields a material that decomposes naturally and avoids the use of chemical treatments. The life cycle of silkworms is preserved, with cocoon formation occurring at approximately 35 days, followed by metamorphosis into moths, reproduction, and natural death within 5 days. This methodology also conserves mulberry trees, the primary source of food for silkworms. A single tree can sustain nearly 100 silkworms and 1 acre can yield approximately 13–14 kg of raw silk. Depending on the silkworm variety and diet, naturally occurring reddish or golden silk hues are highly valued. Furthermore, mulberry trees provide sustainable materials for local artisanal practices such as basketry, furniture production, and traditional medicinal applications [32, 33].

4.2.2 Peace Silk

Peace silk production employs a technique that preserves silkworms by allowing them to complete their life cycles. Silk is harvested after the silkworms transform into moths and vacate their cocoons. The abandoned cocoons were collected and the broken threads were reconstituted by spinning. Prior to spinning the silk thread, each cocoon underwent an individual examination to verify that the moth had exited. Once the moth has emerged, breaking the silk fibre, a degumming procedure is employed to remove sericin. Subsequently, silk is processed in a manner akin to cotton or hemp, and spun rather than wound onto reels as a continuous thread. This approach ensures that no harm occurs to the moths while preserving the eco-friendly qualities of the fabric (Ahimsa/Peace) [31, 34].

4.2.3 Wild Silk

In their natural habitat, silk caterpillars are permitted to thrive without being harvested by the fashion industry. Tussah and Tussur, also known as Wild Silk, are natural and organic products. It is created by silkworms living freely in their natural environment, where they complete their life cycles, transform into moths, mate, and subsequently perish. The production of Wild Silk is a labour-intensive process. Local communities gather cocoons, manually spin the fibres, or weave them into fabric for commercial purposes. This environmentally friendly approach safeguards forests from destruction while providing employment for the locals. Wild silkworms produce silk that has uniqueness in colour and feel when compared to domesticated *Bombyx mori*, with shorter cocoon strands owing to moth emergence [35, 36].

Eri silk from domesticated *Philosamia ricini* was nearly as white as *B. mori* silks. Although cultivated from domesticated worms, Eri, also renowned as peace silk, caterpillars are not killed to complete their life cycle as moths. Eri silk resembles a mix of wool and cotton but retains the softness of silk [37]. Figure 3 (right) depicts eri silk yarn.

4.2.4 Matka Silk

Matka, a distinctive type of Indian silk fabric produced on handlooms, is renowned for its incorporation of robust threads. These threads are uniquely created by utilizing yarns produced from punctured cocoons as weft (peace silk) and organzine yarn as the warp. As yarns are produced from shorter segments of mulberry silkworm silk and are hand-spun with intact sericin, contributing to slubs and irregularities that give the fabric's different aesthetic. Although it bears a resemblance to tweed, Matka silk is distinguished by its uniformly coloured fibres. This fabric is characterized by its lustre and durability while also supporting sustainable silk production practices. In addition, matka silk is biodegradable [36, 38].

4.2.5 Recycled and Vintage Silk

Investing in an antique silk garment that is undamaged and free from permanent marks is considered a wise choice. When properly cared for, silk can endure for thousands of years; archaeological excavations in China have uncovered silk fabrics dating back to more than 4000 years. Vintage silk clothing is ideal for recycling and repurposing purposes. Recycling old silk is a practical choice, as the fabric generally remains intact, making the ideas of reuse and restyling both logical and beneficial [38].

4.3 Sustainable Spider Silk

Spider silk is produced by spiders from an abdominal gland that secretes a viscous protein solution, which is quickly converted into an insoluble solid fibre. This silk, used by spiders to capture prey, exhibits exceptional mechanical properties compared to cultivated mulberry silk [39, 40]. Artificial production uses other organisms owing to the cannibalistic and territorial nature of spiders. Chimeric silkworms produce a combination of endogenous silkworm silk and spider silk. In a study by Nexia Biotechnologies, transgenic goats with spider dragline silk genes inserted into their mammary gland cells produced spider silk proteins in their milk [41]. Injecting spidroins into hosts, such as bacteria, plants, and mammals, is an environmentally favourable approach because of the use of renewable feedstock. However, the chosen host can have environmental impacts, including demands for water and energy, land use, and the necessity for managing waste and intermediate products to prevent potential health risks.

4.4 Vegan Leather

Historically, leather and fur have symbolized luxury and wealth [42]; however, the processes involved in their production have raised significant ethical issues. Despite its condemnation of animal slaughter, the fashion sector has widely utilized leather for clothing and accessories. In the late twentieth century, growing consumer interest in alternatives to leather emerged, driven by an increased awareness of animal welfare [43]. This change has led to a notable increase in the market for synthetic leather, which is produced from fabric substrates coated with materials such as polyurethane or polyvinyl chloride [44, 45]. Conventional methods of leather production pollute groundwater and emit toxic substances, including chromium [46], raising serious environmental concerns associated with petroleum-derived products. As awareness of climate change has intensified, there has been a notable shift in consumer preferences towards sustainable options, leading to the emergence of vegan leather, which is crafted from the pulp of fruits, leaf fibres, and agri-waste combined with different synthetic binders. The vegan movement has profoundly influenced the textile and fashion industries [47] by encouraging established brands to launch

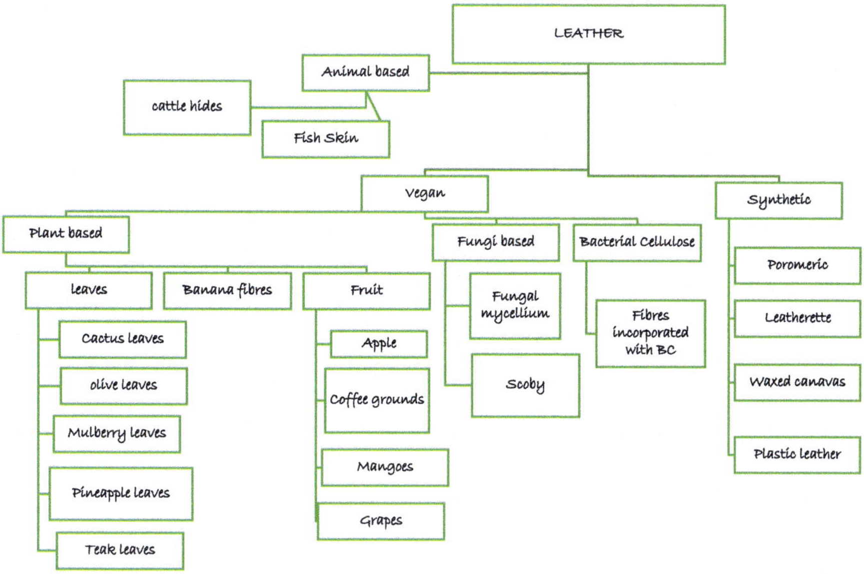

Fig. 4 Classification of leather [51]

vegan product lines [48, 49]. Furthermore, certain bacteria, including species from the genera Azotobacter and Gluconacetobacter, can produce pure cellulose devoid of lignin and hemicellulose [50]. Additionally, fungal mycelia represent a biological and environmentally friendly alternative to traditional leather, with both microfungi and macrofungi used to develop materials that resemble leather. Mushroom-based leather has emerged as a vegan-friendly yet sustainable alternative to animal-derived leather. Figure 4 depicts the different sources from which leather is produced.

5 Ecofriendly Dyeing Process

Wet textile processing is challenging in terms of water consumption. Multifunctional reactive and biodegradable dyes are being developed to reduce water and energy use while eliminating toxicity [52]. Combined with efficient dyeing performance and environmental improvements, they serve as eco-friendly solvents with zero emission [53, 54].

5.1 Pre-treatment

Traditional methods of dyeing cotton necessitate considerable amounts of salt and water, resulting in substantial wastewater production. In response, researchers have developed a cationic fibre modification technique that is more environmentally sustainable and utilizes reduced energy, water, and salt. This innovative approach involves the application of *N*-(3-chlor-2-hydroxypropyl) trimethylammonium chloride to cotton fibres, which endows them with a permanent positive charge that effectively attracts negatively charged dyes, thereby minimizing dye consumption. Dow's ECOFAST Pure pre-treatment process generates cationic cotton with a durable positive charge, achieving a 50% reduction in both water and dye usage. Similarly, ColorZen's pre-treatment employs a solution containing a wetting agent, caustic soda, and ammonium salt, leading to a 95% decrease in toxic chemical usage and a 90% reduction in water waste. Initiating the process in the field, this method enhances dyeing efficiency, involving 90% reduced water usage, 75% decrease in energy consumption, and 90% reduction in chemical usage, while halving dye consumption compared to conventional salt-based techniques. ColorZen has received the OekoTex Standard 100 certification, confirming that it is clear from toxic substances and safe to be utilized in products intended for individuals of all ages [55].

5.2 Plasma Technology

Plasma technology is predominantly utilized to modify surfaces and enhance the properties of textiles, thereby improving dyeing rates, colour quality, diffusion, and dye adhesion. During this process, the textile is positioned within a chamber where plasma activation takes place, leading to the generation of particles that engage with the textile surface. This interaction leads to the development of a nanoscale finer film on the material surface, consisting of a variety of functional groups [56].

5.3 Enzymatic Pre-treatment

Enzymatic applications in the textile sector promote environmentally sustainable preparatory and low-temperature dyeing methods. Enzymes such as pectinases, cellulases, proteases, peroxidases, and lactases effectively compensate for traditional chemical agents in the dyeing process. The application of proteases enhances the dispersion of dye molecules within wool fibres, thereby increasing the adsorption rate by reducing the apparent activation energy of the dye. Similarly, cellulases contribute to the enhancement of colour brightness in cellulosic fabrics [57].

5.4 Air Dyeing

This technology, referred to as water-free dye technology, employs air flow as the medium for dye transportation to textiles, effectively substituting water. This ground-breaking method eliminates the need for subsequent treatment or finishing processes, enabling dye application on both sides of the fabric. The dye effectively permeated the fabric, mitigating problems, such as leaching and fading. In this process, the air is utilized to inject dye into the fibres, and the fabric is heated to facilitate the direct injection of dye in gaseous form. This method presents considerable benefits compared to traditional dyeing techniques, including vat dyeing and cationic dyeing, resulting in more vibrant and durable colours throughout the textile value chain [55, 58, 59].

5.5 Natural Mordants in Natural Dyeing

Mordanting is crucial for obtaining vivid hues when utilising natural dyes, which are sustainable because of their origin from renewable materials. Natural dyes inherently possess a low affinity for fibres, necessitating the use of mordants such as metallic salts, including alum, copper sulphate ($CuSO_4$), chrome, stannous chloride, and ferrous sulphate ($FeSO_4$). Historically, dyers have employed hazardous substances, such as chromium and tin, in addition to alum, copper, iron, and even stale urine. Recent studies have investigated environmentally friendly metal salts, such as alum and iron sulphates, along with tannins and botanical extracts for mordanting purposes. Notable examples include the use of mango bark, staghorn sumac leaves for cotton, rhubarb leaves for animal fibres, juniper needle ashes as an alternative to alum, gallnut extract, myrobalan extract, and natural tannins derived from sources such as black tea, green walnut hulls, oak galls, and pomegranate rinds [55].

5.6 Supercritical Fluid CO_2

Supercritical fluids represent a unique state of matter that possesses the characteristics of both liquids and gases, serving dual roles as solutes and solvents, respectively. These substances exist above their critical temperature and pressure, preventing them from transitioning into gaseous or liquid states. They are characterized by the high diffusivity of solutes and low viscosity, although their transport mechanisms can be intricate. In a carbon dioxide (CO_2) dyeing apparatus, essential components include a CO_2 cylinder, pressure pump, temperature control unit, reaction vessel, and temperature adjustment (heating/cooling) mechanism. This system operated under a pressure of 800 psi, with continuous stirring at 1000 rpm, and was maintained at approximately 180 °C. Subsequently, the pressure was elevated to 3500 psi and sustained for a duration of 2 h. Upon depressurization, the dyed fibres were extracted. This dyeing technique offers several advantages, including the absence of wastewater generation, reduced energy requirements due to the elimination of drying processes, and enhanced dye diffusion facilitated by the swelling effect induced by CO_2 [60, 61].

5.7 Liposomes-Based Technology for Dyeing

Liposomes are synthetic phospholipid vesicles with an aqueous core, with diameters ranging between 10 and 100 nm. These structures contain hydrophilic phosphate, choline groups along with a hydrophobic hydrocarbon component that acts as a dye carrier enabling easy absorption by wool fibres. The interaction between the liposomal lipids and the wool cell membrane complex facilitated uniform dye distribution within the fibre at relatively low temperatures (85 °C). In commercial settings, liposomes are used as textile auxiliaries in the dyeing of wool and have been integrated into various textile production processes. This approach offers numerous benefits, including enhanced smoothness, maintenance of mechanical properties, and reduction of wastewater generated by dyeing facilities [62].

5.8 Microbial-Based Dyes

Microbial-based dyes, especially those produced using enzymes and fungi such as *Basidiomycota* [63] and *Talaromyces purpurogenus* sp. [64], offer cost-effective and energy-saving alternatives. These bio-dyes, typically in liquid form, reduce toxic effluents compared with synthetic dyes, which require large amounts of water.

5.9 Sustainable Colouring of Silk

Textile fibre dyeing is a major polluter, prompting the need for eco-friendly methods. A researcher [65] assessed the environmental impression of silk production and found it to be significantly higher than that of cotton, nylon 66, and wool in terms of agricultural land use, blue water footprint, energy need, eutrophication of fresh water, impact of climate change, and urban land utilization. Below are some ways to obtain sustainably coloured silk [66].

5.9.1 Silk with Natural Colouration Produced by Wild Silkworms

During the larval phase, silkworms primarily feed on mulberry leaves that are abundant in carotenoids. In silkworms, these carotenoid pigments are assimilated by the gut epithelium, transported to the haemolymph, and stored in the labial gland, leading to the development of yellow-coloured cocoons. The carotenoid pigments are combined into sericin, and the removal of sericin gum during the degumming process leads to white silk fibroin production. Selective silkworm breeding methods have been adopted to produce white cocoons, a result associated with a deficiency in carotenoid transfer [66].

5.9.2 Genetically Modified Silk

Numerous genetic engineering methodologies have been analysed to improve the efficacy and stability of the genetic transformation processes. Several research efforts have focused on incorporating fluorescent protein genes into domesticated silkworms with the aim of utilizing these organisms as bioreactors for the production of recombinant proteins [67]. The primary function of fluorescent proteins is to assist as markers rather than being the final product, with cocoons displaying luminescence in green, orange, and red hues.

Researchers have devised an eco-friendly method in which silkworms spin coloured silk cocoons directly after consuming a dye-modified diet. This biological integration of dyes into silk fibroin eliminates the need for additional dyeing, conserving water, energy, and chemicals [68]. Various dyes, including blue dyes such as Berlin blue, Nile blue, methylene blue, and indigo carmine, and others like fuchsin, gentian violet, alizarin, methyl orange, and safranin, have been tested in silk [31].

5.10 Zero Discharge Dyeing

The zero-discharge dyeing technique effectively tackles issues related to water scarcity and pollution by recycling textile wastewater that contains dyes, acids, and salts from diverse industrial sectors. This methodology significantly reduces the ecological footprint of textile effluents. The recycling process entails the treatment of wastewater from the dyeing unit using a jigger for future dyeing operations. This method is characterized by its cost-effectiveness and energy efficiency, while also promoting water conservation and complying with sustainable industrial practices [69].

6 Printing

Rotary screen printing, a longstanding method in textile printing, has undergone significant transformations with the advent of digital printing technology that was initially developed by Dr. Sweet at Stanford University in the 1980s [70]. Digital printing presents numerous sustainable advantages, including decreased water and energy usage, the ability to sample an unlimited range of colours, rapid design alterations, superior reproducibility, and the absence of restrictions on design dimensions [71, 72]. This technology, however, presents certain disadvantages, including reduced printing speed, higher costs associated with ink, and the necessity for specialized pre-treatment of fabrics. Research indicates that pre-treatment with atmospheric pressure plasma technology markedly enhances colour yield, fastness, sharpness, and antibacterial characteristics in digital printing on cotton, with these improvements persisting through several washes [73].

7 Finishing

A range of hazardous substances, including formaldehyde-based softeners, dye fixatives, and resins, have been employed to improve the functional properties of textile materials. The release of free formaldehyde from these chemicals can lead to skin irritation, allergic reactions, and heightened environmental pollution, if not properly managed. It is essential to prioritize sustainability in finishing processes to mitigate the ecological impacts. Recent innovations have resulted in the development of sustainable techniques and materials, such as natural dyes, that not only impart colour to textiles but also offer antimicrobial properties, ultraviolet (UV) protection, and mosquito repellence. Biopolymers sourced from renewable materials, including plant starches, proteins, and cellulose, are viable alternatives to traditional petroleum-derived polymers. These biopolymers can be optimized for their superior mechanical properties, thermal stability, and barrier function, making them applicable in diverse fields [74]. The application of nanotechnology allows for the incorporation of finishes that provide water resistance, UV shielding, and antimicrobial features to fabrics while maintaining biodegradability. Such advancements contribute to prolonging the lifespan of textiles and ensuring their safe reintegration into the environment after use. Furthermore, the inclusion of nanoparticle (NP) fillers within biopolymer matrices enhances the interfacial adhesion, thereby improving the performance of composite materials. This synergy between biopolymers and nanotechnology not only boosts performance but also meets the rising demand for sustainable manufacturing methods.

The integration of plant extracts [75] and essential oils into textiles through microencapsulation or nanoemulsion techniques, as highlighted by Bakry et al. [76], offers a range of functional benefits. These benefits include antimicrobial properties that prevent the growth of microorganisms, UV resistance from compounds that absorb or reflect UV light, long-lasting fragrances [77], insect repellence [78], and antioxidant properties that protect against oxidative damage. Both the microencapsulation and nanoemulsion methods allow for controlled release, enhance the stability of volatile extracts, improve durability, and ensure even distribution. The environmentally friendly and cost-effective green synthesis of NPs, particularly plant-mediated synthesis uses plant extracts as reducing and capping (stabilizing) agents in contrast to conventional chemical synthesis methods. This approach produces NPs including silver, gold, copper oxide, and zinc oxide, which, when incorporated into textiles, boost antimicrobial effectiveness, provide UV protection, enable self-cleaning features, and, in some instances, offer electrical conductivity.

Enzymatic treatments, such as biopolishing, also improve cotton fabrics, whereas papain enzymes reduce wool shrinkage [79, 80]. Table 4 shows some of the green alternatives to conventional solvents.

Table 4 Eco-friendly alternatives for the conventional toxic chemicals [81, 82]

Process	Conventional chemical	Alternative/substitute	Use
Sizing	Starch	Water-soluble polyvinyl alcohol	Yarn sizing
Desizing	Hydrochloric acid	Amylases	Removing size material from fabrics
Scouring of cotton	Sodium hydroxide	Pectinases	Cleaning impurities from cotton
Bleaching	Hypochlorites	H_2O_2	Fabric bleaching
Oxidation of vat/sulphur dyes	Potassium dichromate	H_2O_2 (Hydrogen Peroxide), $NaBO_3 \cdot 4H_2O$ (Sodium perborate)	Oxidation in dyeing
Thickener	Kerosene	Water-based PA (polyacrylate) copolymers	Used as a thickener in printing processes
Hydrotropic agent	$CO(NH_2)_2$ (Urea)	Dicyandiamide (partially)	Dyeing process
Water repellents	$C_8 F_{18}$ (C_8 fluorocarbons)	C6 fluorocarbons	Imparting water-repellent properties
Crease recovery chemicals	CH_2O(Formaldehyde) based resin	Polycarboxylic acids	Improving fabric crease resistance
Wetting agents/detergents	Alkyl phenol ethoxylates	Fatty alcohol phenol ethoxylates	Wetting and cleaning fabrics
Neutralization agent	CH_3COOH (Acetic acid)	Formic acid	pH adjustment after scouring or bleaching
Peroxide killer	$NA_2S_2O_3$ (Sodium thiosulphate)	Catalases	Removing excess hydrogen peroxide after bleaching
Mercerization	NaOH (Sodium Hydroxide)	Liquid ammonia	Strengthening cotton fibers
Reducing agents	Na_2S (Sodium sulphide)	Glucose, acetyl acetone, thiourea dioxide	Dyeing process for sulphur dyes
Dyeing	Powder form of sulphur dyes	Prereduced dyes	Fabric dyeing
Flame retardant	Brominated diphenylethers	Inorganic salts, phosphonates	Flame retardant treatment for fabrics
Shrink proofing	Chlorination	Plasma treatment	Preventing fabric shrinkage
Size preservative	Pentachlorophenol, formaldehyde	Potato starch or CMC	Preserving yarn size
Stain removers	Carbon tetrachloride, sodium hypochlorite	Enzymatic stain removers	Removing stains from fabrics
Peroxide stabilizer	Sodium silicate, phosphorus-based compounds	Nitrogenous stabilizers	Stabilizing hydrogen peroxide in bleaching
Detergent/Emulsifier	Nonylphenol ethoxylates (APEOs), synthetic surfactants	Eco-friendly surfactants from dextrins	Cleaning fabrics and emulsifying oils

(continued)

Table 4 (continued)

Process	Conventional chemical	Alternative/substitute	Use
Softener	Silicones, amino silicones + APEO emulsifiers	Beeswax, aloe vera, Vitamin A, wax emulsions	Imparting softness to fabrics
Coatings/ Degreasing	Kerosene, dichlorobenzene, trichlorobenzene	Solvosurfactants from glycerol	Coating and degreasing textiles
Oxidation in dyeing	Sodium dichromate	Hydrogen peroxide	Oxidation of dyes during dyeing of textiles
Pigment printing	Water-based thickeners	benzoic acid, Butyl benzoate	Fabric printing pigments
Finishing/Dye fixing	Formaldehyde	Polycarboxylic acids, non-formaldehyde products	Fixing dyes on fabrics and enhancing fabric durability
Softener (functional finish)	Silicones, amino silicones + APEO emulsifiers	Environment-friendly softeners, wax emulsions	Enhancing fabric softness and finish

8 Textile Effluent Treatments

The textile industry substantially contributes to global water pollution, thus prompting the exploration of various treatment technologies. Conventional techniques, such as adsorption and coagulation, often fail to effectively eliminate the complex dyes and chemicals present in textile effluents. Recent developments have focused on integrating innovative technologies to improve treatment efficiency and reduce environmental repercussions. Prominent methods include advanced oxidation processes (AOPs), membrane filtration systems, and biological treatments tailored for textile effluents, which collectively demonstrate the potential to eradicate persistent dyes, reduce the chemical oxygen demand (COD), and enhance water quality. Furthermore, researchers are exploring the application of nanotechnology and smart materials to develop sustainable and cost-effective solutions for wastewater treatment. An advanced treatment technique, peroxi-electrocoagulation, combines the process of electrocoagulation with hydrogen peroxide and is effective in diminishing the COD and sulphate concentrations in dye waste waters [83]. A synergistic approach that integrates coagulation, flocculation, and AOPs has shown marked improvements in colour and COD removal, establishing it as a feasible method for treating textile wastewater [84–86]. Furthermore, biochar derived from waste biomass has surfaced as an economical option for adsorbing hazardous dyes and heavy metals, thereby enhancing the treatment of textile effluents [87, 88]. Membrane filtration methods, including nanofiltration and reverse osmosis processes, are gaining traction for the treatment of textile effluents owing to their exceptional ability to eliminate dyes and dissolved solids. These techniques yield high-quality treated water that can be utilized in textile production, thereby contributing to water conservation initiatives. The combination of membrane filtration with other treatment strategies, such as biological processes or advanced oxidation, has also been explored for enhanced effectiveness.

9 Circularity in Textiles

Embracing circularity in the textile industry is vital for achieving sustainability, as the conventional linear model contributes 10% of the global carbon emissions [89, 90]. A circular economy (CE) addresses these issues by promoting recycling, reuse, and sustainable practices [90, 91]. CE in textiles emphasizes designing durable and recyclable products, minimizing waste, and using innovative recycling technologies to reclaim value from end-of-use products [92, 93]. Biotechnological approaches further enhance sustainability by utilizing renewable resources and reducing the use of harmful chemicals [94]. However, supply chain challenges, consumer perceptions, lack of standardized approaches, and complex waste management hinder the circularity of textiles [95–98]. Achieving a CE necessitates collaboration among stakeholders to develop circular business models focusing on sustainable design and resource efficiency [99, 100] as well as educational initiatives to raise awareness and empower consumers. Implementing recycling and upcycling practices, such as fibre regeneration and thermal recovery, along with robust collection and sorting systems, and innovative technologies, such as near-infrared spectroscopy, can significantly mitigate the environmental footprint of textile waste [90, 93, 101].

Eco-design principles that focus on durability, repairability, and recyclability are crucial for circularity [102]. Integrating sustainability into the design and implementation of EPR policies incentivizes circularity [95, 103]. Collaboration among stakeholders is essential for a circular textile economy, enhancing understanding and overcoming "silo mentality" [104]. Consumer education promotes sustainable behaviour and encourages participation in rental, resale, and recycling programs [105]. Policy frameworks, including regulations, incentives, and international collaboration, have driven the transition to a circular textile economy [91, 95, 103, 106]. Innovative recycling, eco-design, stakeholder collaboration, and supportive policies are the foundations for achieving circularity and reducing the environmental load.

Textile and fashion brands are increasingly adopting circular principles for sustainability. "Patagonia" promotes resale and recycling through its Worn-Wear program [107]. "Eileen Fisher's" "Renew" take-back program prioritizes sustainable materials and designs for recycling [108]. "Reformation" uses sustainable materials, promotes transparency with its "RefScale" tool, and offers a take-back program [109]. "H&M's" "Conscious" collection and garment collecting initiative reflect fast fashion brands' circularity efforts [110]. "Levi's" "WellThread" collection emphasises product longevity, responsible consumption, and recycling [111].

10 Life Cycle Assessment (LCA)

LCA is crucial while evaluating the environmental impact of textile production and consumption, promoting sustainability by identifying critical environmental degradation points and informing better practices throughout the entire life cycle [112]. It assesses impacts, such as GHG emissions and resource depletion, especially in blended textiles. LCA supports a CE by evaluating recycling methods, showing that industrial textile recycling offers substantial environmental benefits [113] such as reducing pollution in the silk industry through effective waste recycling [114]. Moreover, LCA addresses social aspects, such as labour rights and community impacts, ensuring a comprehensive approach to the textile industry's sustainability challenges [115, 116].

11 Luxury Fashion and Sustainability

Luxury fashion's dedication to quality, durability, and exclusivity should be grounded in ethical values and the pursuit of excellence, leading affluent consumers to avoid products that cause environmental or social harm. The actions taken by a luxury brand will significantly shape its legacy over the next 50 years as it adapts to contemporary values and future goals. In a landscape marked by scarcity, inequality, and cultural shifts, luxury can have a positive impact within a broader ecological framework by tackling both environmental and social concerns with quality and heritage bolstering sustainability. Achieving sustainability requires collaboration across the supply chain, including cleaner production techniques, responsible sourcing, and conscious consumption, which are all supported by measurable performance indicators. Sustainable luxury seeks to reduce resource consumption, while simultaneously improving product quality and longevity. Suppliers demand loyalty from buyers and sellers, underscoring the importance of communication, trust, and authentic engagement in sustainability initiatives. However, the fashion industry often struggles with weak supplier relationships as buyers tend to focus on cost-cutting, leading to unfair wages and excessive working hours [117]. Subcontracting to factories that fail to meet established standards can happen without the purchasing company's awareness. It is crucial to align ethical strategies and provide incentives to improve working conditions, which requires a comprehensive business model that connects design, procurement, production, and distribution with enhanced ethical practices [118]. Companies should go beyond merely addressing surface-level issues and embed sustainability into core elements such as principles of supply chain design and configuration. Implementing transformative changes requires significant resources to integrate sustainability into all business decisions, potentially leading to better financial results, reduced brand risks, increased operational efficiency, and the development of an ethical market segment. Strong occupational health and safety policies play a key role in identifying and addressing workplace hazards, thus lowering injuries and absenteeism, with upstream suppliers assisting fashion firms in adjusting their inbound logistics. Effective sustainable supply chain management requires establishing internal practices before pursuing external partnerships. Luxury fashion brands are encouraged to cultivate supply chain relationships that prioritize innovation over mere cost considerations, aligning with societal expectations [119]. Strong relationships with stakeholders enhance the understanding of company operations and strategic direction, whereas weaker connections may only provide short-term advantages. A genuine commitment from the management is essential to ensure compliance with the purchasing firm's guidelines. It is vital to integrate sustainability into every phase of the supply chain, especially during the design phase at both the macro and micro levels, to promote the expansion of innovative and sustainable strategies within the luxury fashion industry.

12 Conclusion

The textile industry, particularly the luxury fashion sector, has a significant environmental impact owing to its high consumption of resources, energy, and chemicals. Green chemistry and clean technology offer potential solutions for reducing this impact by focusing on sustainable practices throughout the product lifecycle. Green chemistry principles aim to prevent waste, use less hazardous materials, reduce energy consumption, and promote the utilization of renewable feedstocks by gradually replacing conventional sources. Clean technology complements green chemistry by optimizing processes for recycling, conserving, and using resources efficiently. The luxury fashion industry is compelled to adopt sustainable practices to meet consumer demand and address environmental concerns. By embracing green chemistry and clean technology, the industry can work towards reducing its carbon footprint and ensuring a more sustainable future.

References

1. Essert, V., Belle, S., Millet, L., Etienne, D., Masclaux, H., & Verneaux, V. (2024). Impacts of anthropogenic eutrophication on the carbon transfers in lacustrine food webs: A paleolimnological assessment in four mountain lakes. The Holocene, 34(5), 604–618. https://doi.org/10.1177/09596836231225726
2. Fang, T., Li, H., Zhao, X., Li, J., Hou, G., Yang, K., Lu, W., Liang, Y., & Cui, K. (2019). An overview of heavy metal pollution in Chaohu Lake, China: Enrichment, distribution, speciation, and associated risk under natural and anthropogenic changes. Environmental Science and Pollution Research, 26, 29585–29596. https://doi.org/10.1007/s11356-019-06210-x
3. Clarke, P. (1980). Microbiology and pollution: The biodegradation of natural and synthetic organic compounds. Philosophical Transactions of the Royal Society of London. Series B, Biological Sciences, 290, 355–367. https://doi.org/10.1098/rstb.1980.0100

4. Estimating the carbon footprint of a fabric. (2011). https://oecotextiles.wordpress.com/2011/01/19/estimating-the-carbon-footprint-ofa-fabric/
5. Aiama, D., Carbone, G., Cator, D., Challender, D., & GB, O. (2016). Biodiversity risks and opportunities in the apparel sector. International Union for Conservation of Nature.
6. Aspinall, M. (2022). Back to the future of fashion past: Re-fashioning future garment making. In Fashion heritage (pp. 81–112). Springer.
7. Sobuj, M., Nishat, R., Alam, M., Ahmed, M., Jannat, N., & Akter, M. (2024). Willingness of buyers to buy wearable rejected clothes from clothing manufacturers: Model for fostering sustainable apparel business. Tekstilec, 67(3), 1–11. https://doi.org/10.14502/tekstilec.67.2024045
8. Zhang, Y. (2020). Research on the dilemma of sustainable fashion design. American Journal of Art and Design, 5, 59. https://doi.org/10.11648/j.ajad.20200503.11
9. Fishwick, M. (2012). A carbon footprint for UK clothing and opportunities for savings. WRAP.
10. Jain, S. (2019). Factors affecting sustainable luxury purchase behavior: A conceptual framework. Journal of International Consumer Marketing, 31, 130–146.
11. Yürük-Kayapınar, P., & Ergan, S. (2020). Consumer behaviors and perceptions towards luxury brands. In Building consumer-brand relationship in luxury brand management (pp. 24–44). IGI Global.
12. Nayak, R., Jajpura, L., & Khandual, A. (2023). Traditional fibres for fashion and textiles: Associated problems and future sustainable fibres. In Sustainable fibres for fashion and textile manufacturing (pp. 3–25). Woodhead Publishing.
13. Ali, A., Xiaoling, G., Ali, A., Sherwani, M., & Muneeb, F. M. (2019). Customer motivations for sustainable consumption: Investigating the drivers of purchase behavior for a green-luxury car. Business Strategy and the Environment, 28, 833–846. https://doi.org/10.1002/bse.2284
14. Kunz, J., May, S., & Schmidt, H. J. (2020). Sustainable luxury: Current status and perspectives for future research. Business Research, 5, 1–61.
15. Ozuem, W., Ranfagni, S., Willis, M., Roval, S., & Howell, K. (2021). Exploring customers' responses to online service failure and recovery strategies during COVID-19 pandemic: An actor-network theory perspective. Psychology and Marketing, 38, 1440–1459.
16. Hockerts, K., & Wüstenhagen, R. (2010). Greening Goliaths versus Emerging Davids—theorizing about the role of incumbents and new entrants in sustainable entrepreneurship. Journal of Business Venturing, 25, 481–492.
17. The World Commission on Environment and Development. (1987). Our common future. Oxford University Press.
18. Achabou, M. A., & Dekhili, S. (2013). Luxury and sustainable development: Is there a match? Journal of Business Research, 66, 1896–1903.
19. National Research Council. (2005). Sustainability in the chemical industry: Grand challenges and research needs—A workshop report. National Academies Press.
20. Asif, M. (2021). Green synthesis, green chemistry, and environmental sustainability. Green Chemistry Technology Letters, 7, 18–27. https://doi.org/10.18510/gctl.2021.713
21. Kirchhoff, M. M. (2005). Promoting sustainability through green chemistry. Resources, Conservation and Recycling, 44, 237–243. https://doi.org/10.1016/j.resconrec.2005.01.003
22. Rana, K. K., & Rana, S. (2014). Fundamentals, representative applications and future perspectives of green chemistry: A short review. OALib, 1, 1–16. https://doi.org/10.4236/oalib.1100748
23. Etzkorn, F. A. (2019). Green chemistry: Principles and case studies. Royal Society of Chemistry.
24. Jain, D. S., Awasthi, A., & Gupta, K. A. (2024). Green chemistry: A sustainable path to environmental responsibility and innovation. Asian Journal of Research in Pharmaceutical Sciences, 14(1), 51–55. https://doi.org/10.52711/2231-5659.2024.00008

25. Ahmed, F., & Mondal, M. I. (2021). Introduction to natural fibres and textiles. In Fundamentals of natural fibres and textiles (pp. 1–32). Woodhead Publishing.
26. Senthil Kumar, P., & Saravanan, A. (2019). Organic cotton versus recycled cotton versus sustainable cotton. In Organic cotton: Is it a sustainable solution? (pp. 141–155). Springer.
27. Mageshwaran, V., Satankar, V., Shukla, S. K., & Kairon, M. S. (2019). Current status of organic cotton production. Indian Farming, 69, 09–14.
28. MESH. (n.d.). 2822323-eri silk hand spun & hand woven. Retrieved December 16, 2024, from https://www.mesh.org.in/photo_15440464.html
29. Muezart. Undyed eri silk weaving yarn 60/9 | 1Kg. Retrieved December 16, 2024, from https://www.muezart.com/products/yarn-natural-color-of-eri-silk-100g
30. Debnath, S. (2017). Sustainable production of bast fibres. In Sustainable fibres and textiles (pp. 69–85). Woodhead Publishing.
31. Karthik, T., & Rathinamoorthy, R. (2017). 6—Sustainable silk production. In Sustainable fibres and textiles (pp. 135–170). Elsevier.
32. Franck, R. R. (2001). Silk, mohair, cashmere and other luxury fibres. Woodhead Publishing.
33. Prakash, C., & Matthes, A. (2020). Cruelty-free silk and guilt-free fashion. Springer.
34. Mughilan, R., & Arun Kumar, R. (2024). A study on cruelty faced by silkworms in the production of silk. International Journal of Advanced Research in Science Communication and Technology, 4, 422–433. https://doi.org/10.48175/ijarsct-19062
35. Peigler, R. S. (2020). 19—Wild silks: Their entomological aspects and their textile applications. In Handbook of natural fibres (pp. 715–745). Elsevier.
36. Makhija P. Silk root [Internet]. mint. 2014. Retrieved August 28, 2024 from https://www.livemint.com/Leisure/V9is7nTLM0Qgb3JmVCd7oK/Silk-root.html?facet=amp
37. Ozkavruk Adanir, E., İleri, B., Can, F., & Ulasli, B. (2022). An ethical approach to sericulture: Production of peace silk in Hatay/Turkey. Textile, 22, 20–30. https://doi.org/10.1080/14759756.2022.2036072
38. Karthik, T., Rathinamoorthy, R., & Ganesan, P. (2015). Sustainable luxury natural fibers—Production, properties, and prospects. In M. A. Gardetti & S. S. Muthu (Eds.), Handbook of sustainable luxury textiles and fashion (pp. 59–100). Springer.
39. Hisa, Y., Gnesa, E., Jeffrey, F., Tang, S., & Vierra, C. (2011). Spider silk composites and applications. In J. Cuppoletti (Ed.), Metal, ceramic and polymeric composites for various uses. IntechOpen.
40. Karthik, T., & Subramani, K. (2016). Sustainable bio-polymer fibers—Production, properties and applications. In S. S. Muthu & M. A. Gardetti (Eds.), Sustainable fibres for fashion industry (pp. 109–140). Springer.
41. Gole, R. S., & Kumar, P. (2012). Spider's silk: Investigation of spinning process, web material and its properties. Biological Sciences and Bioengineering, IIT Kanpur, India.
42. Achabou, M. A. (2021). Is animal welfare a central issue for consumers of luxury goods? Natural Resources Forum, 45, 18–36. https://doi.org/10.1111/1477-8947.12211
43. Jung, H. J., Kim, H., & Oh, K. W. (2016). Green leather for ethical consumers in China and Korea: Facilitating ethical consumption with value–belief–attitude logic. Journal of Business Ethics, 135, 483–502. https://doi.org/10.1007/s10551-014-2475-2
44. Lu, S., Feng, Y., Zhang, P., Hong, W., Chen, Y., Fan, H., Yu, D., & Chen, X. (2021). Preparation of flame-retardant polyurethane and its applications in the leather industry. Polymers, 13(11), 1730. https://doi.org/10.3390/polym13111730
45. Sun, Z., Ren, S., Wu, T., Wen, J., Fang, J., & Fan, H. (2023). A self-matting waterborne polyurethane coating for PVC artificial leather. Polymers, 15(1), 127. https://doi.org/10.3390/polym15010127
46. Mavrodin, M., Lăzăroiu, G., & Mocanu, C. R. (2015). Leather carbon footprint. In Proceedings of 4th international conference of thermal equipment, renewable energy and rural development (pp. 459–462).
47. Seo, K., & Suh, S. (2019). A study on the characteristics and social values of vegan fashion in H&M and Zara. Journal of Fashion Business, 23, 86–100.

48. Gibson, J. (2020). Brands make believe: Ethical veganism and labelling in fashion. Queen Mary Journal of Intellectual Property, 10(2), 143–151. https://doi.org/10.4337/qmjip.2020.02.00
49. Reimers, V., Magnuson, B., & Chao, F. (2016). The academic conceptualisation of ethical clothing: Could it account for the attitude behaviour gap? Journal of Fashion Marketing and Management, 20, 383–399.
50. Chawla, P. R., Bajaj, I. B., Survase, S. A., & Singhal, R. S. (2009). Microbial cellulose: Fermentative production and applications. Food Technology Biotechnology, 47(2), 107–124.
51. Tewari, S., Reshamwala, S. M., Bhatt, L., & Kale, R. D. (2024). Vegan leather: A sustainable reality or a marketing gimmick? Environmental Science and Pollution Research, 31, 3361–3375.
52. Gulzar, T., Farooq, T., Kiran, S., Ahmad, I., & Hameed, A. (2019). Green chemistry in the wet processing of textiles. In The impact and prospects of green chemistry for textile technology (pp. 1–20). Woodhead Publishing.
53. Bianchini, R., Cevasco, G., Chiappe, C., Pomelli, C. S., & Douton, M. J. R. (2015). Ionic liquids can significantly improve textile dyeing: An innovative application assuring economic and environmental benefits. ACS Sustainable Chemistry & Engineering, 3, 2303–2308.
54. Brennecke, J. F., & Maginn, E. J. (2001). Ionic liquids: Innovative fluids for chemical processing. AICHE Journal, 47, 2384–2389.
55. Mahapatra, N. N. (2023). Modern textile processing. Abhishek Publications.
56. Lakshmanan, S. O., & Raghavendran, G. (2017). 9—Low water-consumption technologies for textile production. In Sustainable fibres and textiles (pp. 243–265). Elsevier.
57. Forte, P. (2011). Biotechnology in textiles—An opportunity of saving water. Institute for New Technologies.
58. Babu, K. M. (2022). Overview on significant approaches to waterless dyeing technology. CRC Press.
59. Mahmud, I., & Kaiser, S. (2020). Recent progress in waterless textile dyeing. Journal of Textile Science and Engineering, 10, 1–3. https://doi.org/10.37421/2165-8064.2020.10.421
60. Abou Elmaaty, T., & Abd El-Aziz, E. (2017). Supercritical carbon dioxide as a green media in textile dyeing: A review. Textile Research Journal, 88, 1184–1212. https://doi.org/10.1177/0040517517697639
61. Amenaghawon, A. N., Anyalewechi, C. L., Kusuma, H. S., & Mahfud, M. (2019). Applications of supercritical carbon dioxide in textile industry. In Green sustainable process for chemical and environmental engineering and science (pp. 329–346). Elsevier.
62. Ammayappan, L., Jose, S., & Arputha Raj, A. (2016). Sustainable production processes in textile dyeing. In Green fashion (Vol. 1, pp. 185–216). Springer.
63. Sójka-Ledakowicz, J., Olczyk, J., Polak, J., Grąz, M., & Jarosz-Wilkołazka, A. (2015). Dyeing of textile fabrics with bio-dyes. Fibres & Textiles in Eastern Europe, 23(1), 120–126.
64. Verma, H., Rani, A., Gahlot, M., Goel, A., & Sharma, A. K. (2019). Sustainable dyeing of wool fabric with Talaromyces purpurogenus. Journal of Applied and Natural Science, 11, 796–801.
65. Astudillo, M. F., Thalwitz, G., & Vollrath, F. (2014). Life cycle assessment of Indian silk. Journal of Cleaner Production, 81, 158–167.
66. Sangappa, S., Dandin, S. B., Trivedy, K., & Kumar, S. N. (2007). Coloured cocoons to coloured silk. Indian Silk, 22–24.
67. Zhao, Y. P. (2008). SSR based linkage and mapping analysis of C, a yellow cocoon gene in the silkworm, Bombyx mori. Insect Science, 15, 399–404.
68. Tansil, N. C. (2011). The use of molecular fluorescent markers to monitor absorption and distribution of xenobiotics in a silkworm model. Biomaterials, 32, 957.
69. Gopalakrishnan, M., Punitha, V., & Saravanan, D. (2019). Water conservation in textile wet processing. In Water in textiles and fashion (pp. 135–115). Elsevier.
70. Javorsek, D., & Javorsek, A. (2011). Colour management in digital textile printing. Coloration Technology, 127, 235–239.

71. Chavan, R. B. (1996). Technological revolutions in textile printing. Indian Journal of Fibre & Textile Research, 21, 50–56.
72. Uppinal, S. V. (2015). Quality assessment of digital printed cotton and viscose fabrics (Doctoral dissertation). University of Agricultural Science, Dharwad.
73. Kan, C. W., Yuen, C., & Tsoi, W. (2011). Using atmospheric pressure plasma for enhancing the deposition of printing paste on cotton fabric for digital ink-jet printing. Cellulose, 18, 827–883.
74. Rolling, V., & Sadachar, A. (2017). Exploring Millennials' purchase intentions towards sustainable luxury brands. Iowa State University Digital Repository.
75. Singh, P., & Goel, A. (2022). Conservation of textile immemorial: The fading past of Uttarakhand Museums. In Handbook of Museums Textiles (Vol. 1, pp. 267–290).
76. Bakry, A. M., Majeed, H., Ali, B., Liang, L., Abbas, S., Abouelwafa, M. Y., & Mousa, A. (2015). Microencapsulation of oils: A comprehensive review of benefits, techniques, and applications. Comprehensive Reviews in Food Science and Food Safety, 15, 143–182. https://doi.org/10.1111/1541-4337.12179
77. Miró Specos, M. M., Escobar, G., Hermida, L., Defain Tesoriero, M. V., Puggia, C., & Marino, P. (2010). Aroma finishing of cotton fabrics by means of microencapsulation techniques. Journal of Industrial Textiles, 40, 13–32. https://doi.org/10.1177/1528083709350184
78. Pavan, M., Rani, A., Jhang, T., & Singh, S. P. (2023). Developing herbal mosquito repellent cotton fabric using the optimized process variables for the safe environment. Materials Today Proceedings, 78, 900–906.
79. Li, K., Li, Z., Zhang, J., Wang, J., Yu, Y., Zhou, M., Wang, Q., & Wang, P. (2024). Eco-friendly shrink-resist finishing of wool through synergistic effect of disulfide bond reducing agent and papain. International Journal of Biological Macromolecules, 273, 133156. https://doi.org/10.1016/j.ijbiomac.2024.133156
80. Mojsov, K. (2017). Enzymatic treatment of wool fabrics—Opportunity of the improvement on some physical and chemical properties of the fabrics. Journal of the Textile Institute, 108, 1136–1143. https://doi.org/10.1080/00405000.2016.1222856
81. Arputharaj, A., Raja, A. S. M., & Saxena, S. (2016). Developments in sustainable chemical processing of textiles. In S. Muthu & M. Gardetti (Eds.), Green fashion: Environmental footprints and eco-design of products and processes. Springer.
82. Roy Choudhury, A. K. (2013). Green chemistry and the textile industry. Textile Progress, 45, 3–143. https://doi.org/10.1080/00405167.2013.807601
83. Gunalan, S. (2023). Treatment of textile dye effluent using peroxi-electrocoagulation. Interantional Journal of Scientific Resear in Engineering Management, 7. https://doi.org/10.55041/ijsrem24529
84. Harrelkas, F., Azizi, A., Yaacoubi, A., Benhammou, A., & Pons, M. N. (2008). Treatment of textile dye effluents using coagulation–flocculation coupled with membrane processes or adsorption on powdered activated carbon. *Desalination, 235*, 330–339. https://doi.org/10.1016/j.desal.2008.02.012
85. Xiao, X., Shen, H., Zheng, H., Sun, Y., Xu, Y., Sun, W., Liu, C., Zhao, J., & Wu, H. (2016). Advanced treatment of actual textile dye wastewater by Fenton-flocculation process. Canadian Journal of Chemical Engineering, 95, 1245–1252. https://doi.org/10.1002/cjce.22752
86. Zayas Péerez, T., Geissler, G., & Hernandez, F. (2007). Chemical oxygen demand reduction in coffee wastewater through chemical flocculation and advanced oxidation processes. Journal of Environmental Sciences, 19, 300–305. https://doi.org/10.1016/s1001-0742(07)60049-7
87. Lin, S.-L., Zhang, H., Chen, W.-H., Song, M., & Kwon, E. E. (2023). Low-temperature biochar production from torrefaction for wastewater treatment: A review. Bioresource Technology, 387, 129588. https://doi.org/10.1016/j.biortech.2023.129588
88. Sontakke, A. D., Mondal, P., Bhattacharjee, A., Purkait, M. K., & Tiwari, S. (2022). Recent advances in biochar production and its applications toward textile industry effluent treatment. CRC Press.

89. Furferi, R. (2021). Special issue on innovative textiles in the era of circular economy. Applied Sciences, 11, 4161. https://doi.org/10.3390/app11094161
90. Valtere, M., Bezrucko, T., & Blumberga, D. (2023). Analysis of textile circularity potential. Environmental and Climate Technologies, 27, 220–232. https://doi.org/10.2478/rtuect-2023-0017
91. Marinova, V. (2023). Sustainability of the European textile industry. E3s Web of Conferences, 452, 05020. https://doi.org/10.1051/e3sconf/202345205020
92. Cao, H., Cobb, K., Yatvitskiy, M., Wolfe, M., & Shen, H. (2022). Textile and product development from end-of-use cotton apparel: A study to reclaim value from waste. Sustainability, 14, 8553. https://doi.org/10.3390/su14148553
93. Juanga-Labayen, J., Labayen, I., & Yuan, Q. (2022). A review on textile recycling practices and challenges. Textiles, 2, 174–188. https://doi.org/10.3390/textiles2010010
94. Rahman, M., Billah, M., Hack-Polay, D., & Alam, A. (2020). The use of biotechnologies in textile processing and environmental sustainability: An emerging market context. Technological Forecasting and Social Change, 159, 120204. https://doi.org/10.1016/j.techfore.2020.120204
95. Chen, W., Nalluri, V., Hung, H., Chang, M., & Lin, C. (2021). Apply dematel to analyzing key barriers to implementing the circular economy: An application for the textile sector. Applied Sciences, 11, 3335. https://doi.org/10.3390/app11083335
96. Kazançoğlu, İ., Kazançoğlu, Y., Yarımoğlu, E., & Kahraman, A. (2020). A conceptual framework for barriers of circular supply chains for sustainability in the textile industry. Sustainable Development, 28, 1477–1492. https://doi.org/10.1002/sd.2100
97. Koszewska, M. (2018). Circular economy—Challenges for the textile and clothing industry. Autex Research Journal, 18(4), 337–347. https://doi.org/10.1515/aut-2018-0023
98. Wójcik-Karpacz, A., Karpacz, J., Brzeziński, P., Pietruszka-Ortyl, A., & Ziębicki, B. (2023). Barriers and drivers for changes in circular business models in a textile recycling sector: Results of qualitative empirical research. Energies, 16, 490. https://doi.org/10.3390/en16010490
99. Daňo, F., Drábik, P., & Hanuláková, E. (2020). Circular business models in textiles and apparel sector in Slovakia. Central European Business Review, 9, 1–19. https://doi.org/10.18267/j.cebr.226
100. Ki, C., Park, S., & Ha-Brookshire, J. (2020). Toward a circular economy: Understanding consumers' moral stance on corporations' and individuals' responsibilities in creating a circular fashion economy. Business Strategy and the Environment, 30(2), 1121–1135. https://doi.org/10.1002/bse.2675
101. Bonifazi, G., Gasbarrone, R., Palmieri, R., & Serranti, S. (2022). End-of-life textile recognition in a circular economy perspective: A methodological approach based on near infrared spectroscopy. *Sustainability, 14*, 10249. https://doi.org/10.3390/su141610249
102. Silobrit, I. (2023). Assessing circular textile industry development. *Economia della Cultura, 20*, 55–67. https://doi.org/10.2478/jec-2023-0005
103. Bukhari, M., Carrasco-Gallego, R., & Ponce-Cueto, E. (2018). Developing a national programme for textiles and clothing recovery. *Waste Management Research, 36*, 321–331. https://doi.org/10.1177/0734242x18759190
104. Staicu, D., & Pop, O. (2018). Mapping the interactions between the stakeholders of the circular economy ecosystem applied to the textile and apparel sector in Romania. *Management and Marketing, 13*, 1190–1209. https://doi.org/10.2478/mmcks-2018-0031
105. Filho, W., Dinis, M., Paço, A., Herédia-Colaço, V., Ávila, L., Dennis, K., & Liakh, O. (2022). Covid-19 and sustainability in textile, apparel and fashion use: An assessment of trends. *Textile Research Journal, 93*, 674–690. https://doi.org/10.1177/00405175221114167
106. Lawrence, J., Bettinetti, R., Torretta, V., & Rada, E. (2023). Investigating the effectiveness of clothes recovery programs in promoting a circular economy: A review. *E3s Web of Conferences, 408*, 01002. https://doi.org/10.1051/e3sconf/202340801002

107. Vehmas, K., Raudaskoski, A., Heikkilä, P., Harlin, A., & Mensonen, A. (2018). Consumer attitudes and communication in circular fashion. *Journal of Fashion Marketing and Management, 22*, 286–300. https://doi.org/10.1108/jfmm-08-2017-0079
108. Piller, L. (2022). Designing for circularity: Sustainable pathways for Australian fashion small to medium enterprises. *Journal of Fashion Marketing and Management, 27*, 287–310. https://doi.org/10.1108/jfmm-09-2021-0220
109. Trejo, H. (2024). Sustaining natural dye plants with post-consumer textile waste. *AATCC Journal of Research, 11*, 194–207. https://doi.org/10.1177/24723444241230120
110. Ruiz-Alba, J., Mediano, J., Ayestarán, R., & López-Tenorio, P. (2022). Fashion events and E-WOM. *Event Management, 26*, 405–417. https://doi.org/10.3727/152599521x16192004803511
111. Zhang, B., Zhang, Y., & Zhou, P. (2021). Consumer attitude towards sustainability of fast fashion products in the UK. *Sustainability, 13*, 1646. https://doi.org/10.3390/su13041646
112. Muthu, S. (2015). *Handbook of life cycle assessment (LCA) of textiles and clothing*. Elsevier.
113. Periyasamy, A. P., Wiener, J., & Militky, J. (2017). 4—Life-cycle assessment of denim. In *Sustainability in Denim* (pp. 83–110). Elsevier.
114. Wu, G. (2022). The silk industry around Naupaktos and its implications. *Byzantinische Z, 115*, 351–366. https://doi.org/10.1515/bz-2022-0013
115. Edirisinghe, L. G. L. M., Alwis, A. A. P., & Wijayasundara, M. (2024). Sustainable circular practices in the textile product life cycle: A comprehensive approach to environmental impact mitigation. *Environmental Challenges, 16*, 100985. https://doi.org/10.1016/j.envc.2024.100985
116. Fatma, Ş., Fidan, E., Kızılkaya, A., & Niğmet, U. (2024). Comprehensive analysis of social subcategories throughout life cycle assessment approach for the textile industry. *International Journal of Life Cycle Assessment*. https://doi.org/10.1007/s11367-024-02340-8
117. Wu, H. J., Su, J., & Hodges, N. N. (2020). Exploring the impact of apparel suppliers' affective and cognitive satisfaction on the buyer-supplier relationship: A power-dependency theory perspective. *International Textile and Apparel Association Annual Conference Proceedings, 77*(1). https://doi.org/10.31274/itaa.11858
118. Quinn, S. (2008). Environmental stewardship and sustainable sourcing. In C. Ulasewicz & J. Hethorn (Eds.), *Sustainable fashion: Why now? A conversation about issues, practices, and possibilities* (pp. 358–385). Fairchild Books.
119. Caniato, F., Caridi, M., Castelli, C. M., & Golini, R. (2009). A contingency approach for SC strategy in the Italian luxury industry: Do consolidated models fit? *International Journal of Production Economics, 120*, 176–189. https://doi.org/10.1016/j.ijpe.2008.07.027

Eco-Production of Apparel: Energy- and Water-Efficient Luxury Textiles

Pavan M, Lata Samant, T. Nageshkumar, Sauda Samreen, and N. G. Yashaswini

1 Introduction

The high-end fashion industry is a vital section of the global economy and has an enormous impact on the environment through its significant energy and water consumption. Textile manufacturing processes, mainly wet processing and finishing, depend heavily on electricity and fuels, contributing to greenhouse gas emissions and environmental pollution [1]. Moreover, water-intensive practices in dyeing, finishing, and processing aggravate water scarcity and lead to the discharge of contaminated effluents into ecosystems [2].

Developing technologies play a transformative role in addressing these challenges, bidding innovative solutions to reduce energy and water consumption while improving environmental sustainability. Advances in waterless dyeing techniques, such as air-dyeing and digital printing, are considerably reducing the industry's dependence on water, producing high-quality fabrics with slight environmental impact [3]. Closed-loop water systems permit filtration and reuse within production cycles, conserving water resources and minimizing chemical release into natural habitats. Similarly, wastewater recycling technologies are revolutionizing water

Pavan M (✉) · S. Samreen · N. G. Yashaswini
NITTE School of Fashion Technology and Interior Design, NITTE Deemed to be University, Bengaluru, Karnataka, India
e-mail: pavan.m@nitte.edu.in; sauda.samreen@nitte.edu.in; yashaswini.ng@nitte.edu.in

L. Samant
National Institute of Fashion Technology, Daman, Daman and Diu, India

T. Nageshkumar
Quality Evaluation and Improvement Division, ICAR—National Institute of Natural Fibre Engineering and Technology, Kolkata, West Bengal, India

Sadhna et al. (eds.), *Crafting Sustainability in Luxury Textiles for a Zero-Waste Future*, Sustainable Textiles: Production, Processing, Manufacturing & Chemistry, https://doi.org/10.1007/978-3-031-86773-6_13

management, allowing factories to purify and reuse water while complying with stringent environmental regulations [4].

As the fashion industry transitions towards sustainability, the adoption of energy- and water-efficient technologies is crucial. These solutions mitigate environmental harm and align with the growing consumer demand for eco-friendly products. Hence, this chapter chiefly focuses on energy- and water-efficient technologies, discussed across various sections.

1.1 The Production Process of Textile Substrates

The textile production process, as depicted in Fig. 1, involves a detailed progression from raw materials to consumer-ready products, showcasing the extensive transformation a textile material undergoes. Energy utilization and water consumption, particularly in wet processing, are essential throughout this process. The major stages include spinning, where raw fibres are converted into yarn; weaving or knitting, which transforms the yarn into fabric; dyeing, where colour is imparted to the fabric using water and chemicals; and finishing, which enhances the fabric's properties such as softness, durability, or water repellence. Each of these steps demands significant energy and resource inputs, reflecting the complexity and resource-intensive nature of textile production.

The textile sector is renowned for its significant contributions to the economy but is equally highlighted for its substantial environmental and social challenges in achieving long-term sustainability. This industry presents numerous environmental concerns, primarily due to its high consumption of energy and emission of pollutants. A staggering volume of over three trillion gallons of fresh water is consumed annually worldwide for fabric manufacturing. Moreover, textile treatment and dyeing activities are responsible for releasing nearly one-third of all chemicals discharged into the environment [6, 7].

Energy consumption is a critical challenge in the textile industry, which comprises numerous energy-intensive plants. Figure 2 shows the standard breakdown of electricity and thermal energy use in the composite textile industry. Electricity is predominant in processes like spun yarn spinning, while wet processing relies more on fuel energy. Energy is one of the most important aspects of the textile industry every part of its supply chain is energy-intensive, and the industry is one of the highest energy consumers [9, 10]. In 2019, an average of 7602.88 tonnes of dyed fabric was produced across 15 textile dyeing mills. This production process required 17,689.43 MWh of electricity and 961.26 million litres of groundwater while discharging 640.24 million litres of treated effluent wastewater into the environment [11]. Scarlett Buckley [12] reported that the energy consumption in textile production is as follows: 23% for weaving, 34% for spinning, 38% for chemical processing, and 5% for miscellaneous activities.

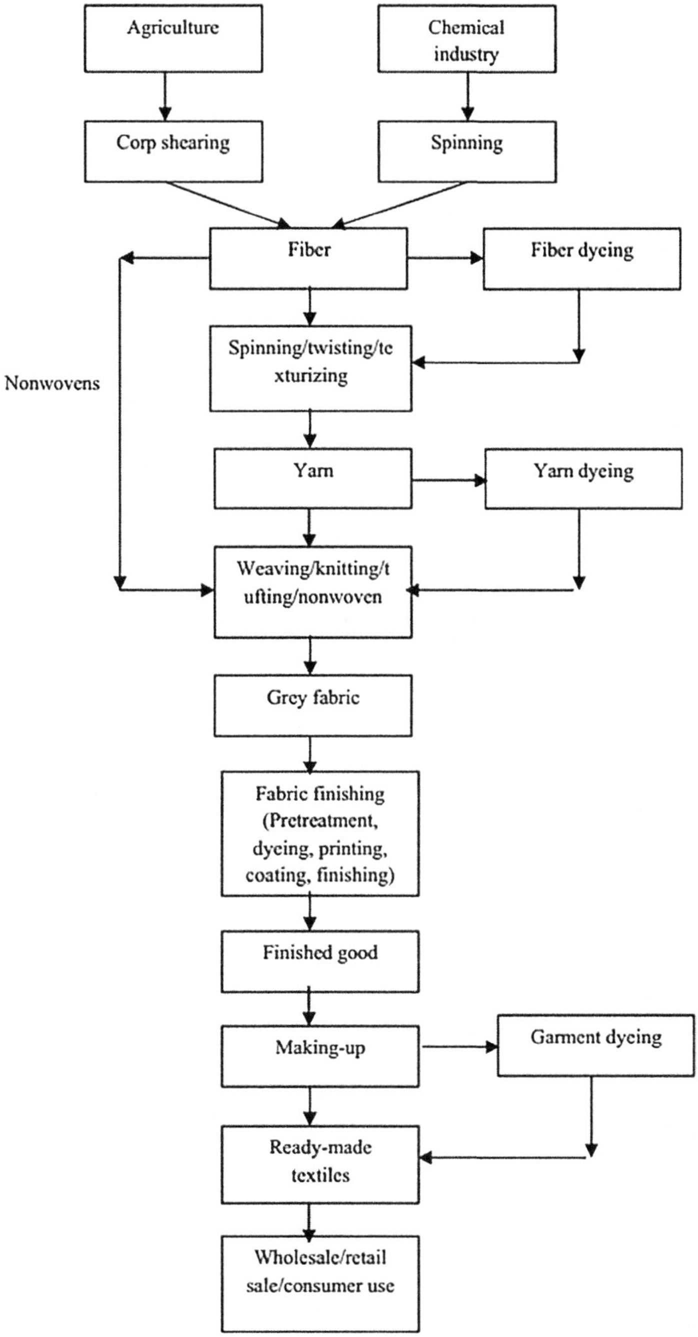

Fig. 1 The textile chain [5]

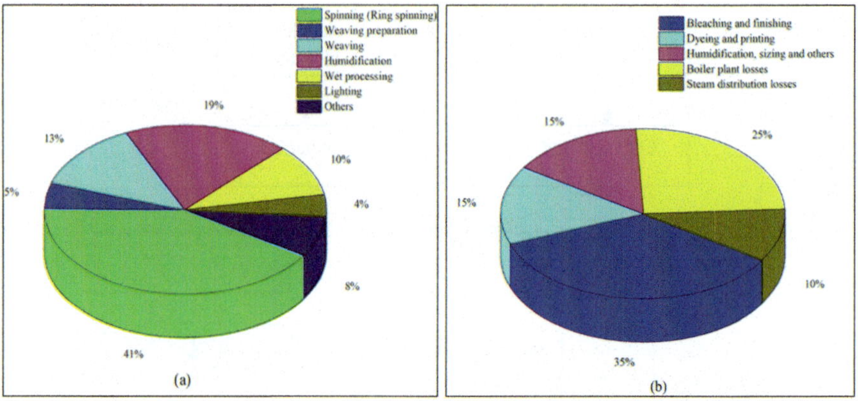

Fig. 2 (**a**) Standard usage of electricity energy. (**b**) Standard usage of thermal energy [8]

2 Sustainable Practices in Luxury Apparel Production

Sustainable production methods for luxury clothing seek to establish a balance between superior craftsmanship and environmental responsibility. In order to ensure ethical sourcing and fair labour methods, luxury businesses are increasingly adopting eco-friendly materials such as organic cotton, recycled fabrics, and bio-based textiles. Advanced technologies, such as 3D printing, zero-waste pattern cutting, and renewable energy, are used to reduce waste and carbon footprints. Sustainability is further promoted through circular fashion schemes, like recyclable designs and repair or resale schemes. These schemes determine how luxury fashion can complement environmental awareness while encountering the growing demand for fashionable and ethical goods.

2.1 Sustainable Sourcing of Materials

Materials that are rare or challenging to obtain due to difficulty in cultivation or the need for trade have long been symbols of wealth and status [13]. The sustainable sourcing of these materials is gaining increasing importance in the luxury apparel sector, as brands recognize the need to align with consumer values and mend their reputation. Luxury brands are retorting to the growing demand for eco-friendly practices by incorporating sustainable materials into their collections. At the same time, they face the challenge of balancing environmental responsibility with the need to maintain the exclusivity and high quality that define luxury.

Sustainable sourcing in luxury apparel focuses on the use of eco-friendly materials shown in Fig. 3. While luxury brands strive to meet the growing consumer demand for environmentally conscious products, they face the challenge of

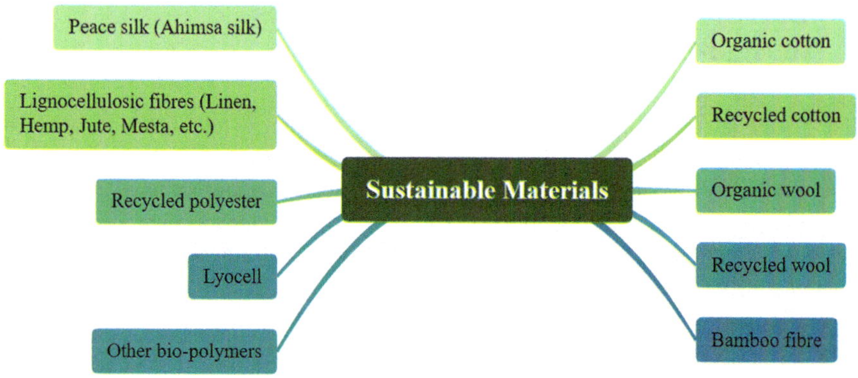

Fig. 3 Some of the sustainable materials for luxury textile production

preserving exclusivity as they incorporate these sustainable practices. Balancing responsible sourcing with the need for premium quality and uniqueness remains a key consideration in the industry [14].

2.2 *Eco-Friendly Dyeing and Finishing Processes*

Eco-friendly dyeing and finishing processes are indispensable to address the environmental impact of the textile industry. Recent advancements focus on sustainable techniques such as natural dye extraction, microbial dyeing, and digital textile printing using biodegradable inks. Innovations such as plasma technology and improved synthetic dyeing methods also augment fabric quality while reducing water, energy, and chemical usage. These developments determine the industry's assurance of balancing functionality and environmental responsibility, flagging the way for a more sustainable future [15, 16, 17].

2.3 *Reducing Waste Through Design and Production*

Reducing waste in luxury apparel production is crucial to promoting sustainability, and innovative design approaches play a crucial role in accomplishing this. Techniques such as zero-waste pattern-making confirm maximum fabric utilization, eliminating leftover materials and minimizing post-production waste. Advanced production methods, coupled with consumer involvement in customization and pre-order models, further support waste reduction. These strategies not only boost sustainability within the fashion industry but also line up with the growing demand for environmentally cognizant luxury products [18, 19].

3 Energy- and Water-Efficient Technologies in High-End Fashion

The textile sector is one of the largest consumers of energy and water, particularly in the production of luxury goods. Given the increasing focus on sustainability and environmental responsibility, there is a pressing need to adopt innovative and sustainable technologies in this domain. The following sections explore energy- and water-efficient technologies along with advanced textile finishes that contribute to sustainability in high-end fashion.

3.1 Energy-Efficient Production Methods

The textile industry is increasingly focused on adopting energy-efficient technologies and solutions to reduce costs and environmental impact. The following section details some key advancements in modern compressors and broader energy-efficient strategies [10, 20–22]:

3.1.1 Innovations in Modern Compressors

Modern compressors have integrated several technological advancements that significantly enhance their energy efficiency in the textile industry. These include the following:

- **Variable Speed Drives (VSDs)**: VSDs allow compressors to adjust their speed and output based on real-time demand. This indicates that, during periods of low demand, compressors can operate at reduced speeds, considerably reducing unnecessary energy consumption. This compliance not only conserves energy but also steers to substantial cost savings.
- **Smart controls and Internet of Things (IoT) integration**: Various latest compressors are embedded with smart controls and IoT integration. These technologies aid in real-time monitoring and optimization of compressor performance, predicting maintenance needs, and preventing downtime. These systems ensure that compressors work at peak efficiency, enhancing overall productivity.
- **Energy recovery systems**: Latest compressors often include energy recovery systems designed to capture and reuse waste heat generated during the compression process. This recovered energy can be employed for heating water or spaces within the facility, further decreasing the overall energy footprint.
- **Improved air filtration and drying**: Efficient air filtration and drying systems incorporated into modern compressors confirm that the air used in production is clean and dry. This not only improves the quality of textile products but also reduces the energy required for subsequent processing stages.

- **Minimized oil fumes and enhanced oil capture**: Efficient air-oil separation systems minimize oil carryover to distribution lines, ensuring cleaner operation. In addition, initiatives to collect contaminated oil from pipelines align with green initiatives, contributing to environmental sustainability.

3.1.2 Fuel Efficiency

As urban air quality deteriorates, preventing air pollution has become critical. One important factor in this effort is the selection of fuels with high calorific values and optimal combustion characteristics. Transitioning from traditional smoke-tube boilers to more efficient water-tube boilers can significantly enhance energy efficiency in the textile industry. Modern boiler technologies improve combustion efficiency and reduce emissions [23].

Heat loss occurs during steam transportation due to pipe radiation, leading to pressure drops. Using smaller-diameter, high-pressure pipes is more efficient for long-distance steam transport than larger, low-pressure alternatives. Effective heat exchangers facilitate indirect heating and cooling of liquids and gases without chemical contamination. These devices play a crucial role in optimizing energy use across various processes within the textile industry [24]. The textile sector is gradually shifting from conventional energy sources to renewable alternatives, like solar, geothermal, wind, tidal, and biomass energy. Solar panels have evolved significantly; now, they are capable of providing robust energy solutions for production units [25].

3.1.3 Energy Auditing for Efficiency Progression

Organizing regular energy audits is crucial for identifying inadequacies and optimizing energy use within the textile industry. Energy auditing includes analysing the flow of energy inputs and outputs to pinpoint areas for improvement. Audits should target visible inadequacies, such as loose electrical connections, mechanical misalignments, as well as inadequate lubrication. Tackling these issues through regular maintenance can lead to considerable reductions in energy consumption [3].

Improving energy efficiency in the textile industry demands optimizing processes, replacing outdated and worn-out machines with advanced technologies and machinery, and adopting post-installation measures to boost benefits such as reduced water and material consumption, higher product quality, and minimized waste. Enhancing electricity efficiency involves optimizing lighting systems, transitioning to modern machines with multiple motor controls, and replacing electric heating with substitutes like steam or gas. Fuel efficiency can be augmented by using high-performance, low-pollution fuels and efficient boilers equipped with smoke extractors and water-pipe systems. Steam systems should integrate high-pressure pipelines and heat exchangers for better energy storage and distribution. Regular energy audits are essential to identify inefficiencies, prevent equipment problems like overheating or friction, and ensure optimal energy transfer. The adoption of sustainable materials, including nanomaterials and synthetics, should focus on environmental responsibility across production and disposal stages. To attain decarbonization and net-zero emissions, approaches such as energy efficiency, industrial electrification, low-carbon fuels, and carbon capture must be implemented, with energy efficiency serving as a cost-effective and impactful means to reduce greenhouse gas emissions [21, 26].

3.2 Water Conservation Practices

Evolving textile manufacturing and wet-processing technologies are changing the textile industry by offering significant energy, environmental, and water efficiency benefits than conventional methods.

3.2.1 Enzymatic Treatments

Employing hazardous chemicals in textile wet processing notably increases biological oxygen demand (BOD), chemical oxygen demand (COD), and total dissolved solids (TDS) in wastewater, further leading to higher treatment costs and environmental pollution. Enzymatic solutions offer an effective substitute to mitigate these challenges. Specific enzyme combinations are utilized for scouring processes, where they act synergistically under mild conditions and across a variable pH range. This approach not only lessens the consumption of energy, water, and chemicals but also reduces overall processing costs. Enzymatic treatments address environmental concerns by reducing the dependency on conventional chemical-intensive methods. By lowering the consumption of chemicals, water, and energy, enzymes contribute to reduced waste generation [27]. Hasan et al. [28] and Mojsov [29] emphasized the role of enzymes in promoting sustainable manufacturing practices in the textile sector.

Recent studies underscore the significant impact of biotechnological advancements in this industry. A study by Chen et al. [30] showed that biotechnological applications, such as bio-clean-up for bleaching, can minimize water and energy usage by approximately 9–14% and 17–18%, respectively. Bio-scouring methods have been shown to reduce water consumption by 30–50%, while bio-stone washing has reduced costs related to water use and air emissions by 50–60% in textile processing facilities in China.

3.2.2 Ultrasound Technology

Ultrasound technology offers a promising approach to reduce the significant use of chemicals, water, and energy in textile wet processes. It positively influences key operations like the de-oiling process, leading to a notable decrease in water and energy needs. The use of ultrasonic waves has been thoroughly investigated in dyeing processes, including reactive and acid dyes, as well as dye dispersants, applicable to almost all types of fibres [31].

Ultrasound has proven to be effective in various fabric preparation processes, including desizing, scouring, bleaching, and mercerization. It has also shown significant potential in auxiliary processes like washing. According to Ramachandran et al. [32], ultrasonic methods can enhance efficiency and lower resource consumption, positioning them as a viable alternative for sustainable textile manufacturing. This technology offers a more sustainable option in textile wet processing by minimizing the use of water, energy, chemicals, and processing time. It can be applied directly with ultrasonic probes or indirectly through ultrasonic baths. This method not only improves efficiency but also reduces environmental impact and enhances fabric quality, making it a promising innovation in the textile industry [33].

3.2.3 Ozone Bleaching of Textiles

Pre-treatment plays a vital role in achieving effective textile dyeing. For darker shades, scouring is usually adequate, but lighter and brighter shades often need extra bleaching. Ozone bleaching presents a sustainable alternative to traditional methods, using significantly less water to achieve desired whiteness and absorbency levels. This process involves exposing pre-wetted fabric with 80% wet pick-up to a mixture of ozone and oxygen in a sealed chamber. Despite its efficiency and eco-friendliness, the setup is costly and requires airtight chambers. A novel enhancement combines ozone bleaching with ultrasonically assisted rinsing, where pre-wetted cotton fabric (50% wet pick-up at pH 7) undergoes ozone treatment for 15 min, followed by a 5-min chemical-free rinse at 40 °C. This approach minimizes resource usage while maintaining fabric quality [34]. A study conducted by Eren and Ozturk [35] highlighted the effectiveness of ozone (O_3) in preparing de-sized and scoured 100% cotton fabrics. The findings indicated that cotton samples bleached with hydrogen peroxide and then treated with ozone for 60 min, as well as those treated for 90 min, achieved comparable levels of whiteness. Notably, the extended ozonation resulted in minimal strength loss in the fabrics, underscoring ozone's potential as a sustainable and efficient alternative to conventional bleaching methods. This approach promotes eco-friendly textile processing by reducing chemical and water usage while preserving fabric quality.

3.2.4 Supercritical Fluid Dyeing of Textiles

Supercritical fluid dyeing (SFD), which mainly uses supercritical carbon dioxide (SC-CO_2), represents a ground-breaking advancement in sustainable textile processing by completely removing water from the dyeing process [36]. SC-CO_2 has distinct properties when it reaches its critical point, merging liquid-like density and solvating ability with gas-like rapid diffusion and low viscosity. This combination allows for efficient dyeing with shorter cycles, lower temperatures, and reduced chemical use, significantly cutting down on energy consumption and emissions [37]. Disperse dyes, which are appropriate for SC-CO_2 dyeing of synthetic fibres like Polyethylene Terephthalate (PET), can be categorized as either temperature or density controllable, enabling precise optimization of dye uptake on substrates. However, the low permittivity of SC-CO_2 limits its effectiveness with hydrophilic dyes, restricting its application to synthetic fibres. Innovative methods, such as employing perfluoropolyether reverse micelles to form micro water pools in SC-CO_2, have made it possible to dye natural fibres like wool with acid dyes at lower temperatures and without additional chemicals. Despite its challenges, SC-CO_2 dyeing offers significant promise as a cost-effective, recyclable, and environmentally friendly alternative to traditional dyeing techniques [37–39].

Figure 4 illustrates various strategies for water conservation in the textile industry, categorized into four main approaches. First, reusing and treating textile wastewater includes methods such as managing effluent complexities, ozonation, filtration, and ultrasound techniques to recycle water. Second, advancements in textile processing machinery, such as low liquor ratio devices, ultrasonic energy, SFD, and hydrojet technology, reduce water usage. Third, innovations in textile manufacturing techniques, like enzymatic processes and cationization, improve water efficiency and Rapid Enzymatic Single-bath Treatment (REST) dyeing. Finally, advancements in auxiliaries and chemicals focus on analysing water footprints, optimal management, cleaner manufacturing practices, and compliance with Integrated Pollution Prevention and Control directives for sustainable water use. These methods collectively aim to enhance sustainability in textile production [40].

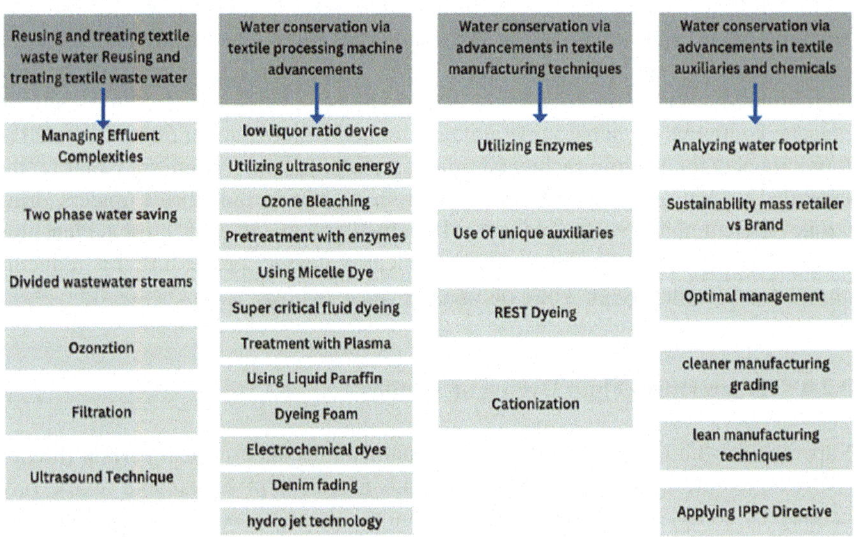

Fig. 4 Strategies for conserving water in the textile industry [40]

3.2.5 Sustainable Retting Methods for Luxury Textile Fibres

Retting, the process of separating fibres from the non-fibrous parts of plant stems, is essential for producing sustainable raw materials like jute and flax used in luxury textiles. Traditional retting methods, which involve submerging plants in water for fibre extraction, are water- and energy-intensive. Studies report that conventional retting accounts for up to 21% of total energy input in jute production (808 MJ/ha) [41]. To address these challenges, advancements such as microbial retting [42], bacterial consortia [43], and enzymatic and chemical accelerators [44] have been developed to reduce water usage, although energy savings remain limited.

Innovations like NINFET-Sathi®, a microbial consortium by ICAR-NINFET, Kolkata, and CRIJAF-SONA, Barrackpur, developed by ICAR-CRIJAF, have successfully cut water consumption by 50% [42, 44]. Ribbon retting, which can be done either manually or mechanically, has shown a remarkable reduction in both water and energy usage when compared to traditional methods. The use of mechanized ribbon extraction further boosts efficiency, leading to lower water use and energy consumption, making it a more sustainable choice for producing luxury textiles. These developments underscore the industry's dedication to environmentally friendly practices in fibre processing [45]. Table 1 illustrates the progress made in energy-efficient mechanical extractors for lignocellulosic fibre retting. These innovations mark important strides in sustainable retting practices, with a focus on minimizing water and energy use.

3.3 *Innovations in Textile Treatment Processes*

The luxury textile industry is experiencing a significant shift thanks to the integration of sustainable and innovative treatment technologies. These developments aim to improve fabric quality, lessen environmental impact, and cater to the increasing consumer demand for eco-friendly luxury products. In contrast to traditional methods that heavily depend on water, energy, and chemicals, contemporary techniques such as air dyeing, foam dyeing, SC-CO_2 dyeing, plasma treatment, and enzymatic processing greatly reduce resource usage while achieving outstanding results. Furthermore, technologies such as laser treatment, ECOFAST Pure, and digital

Table 1 Energy-efficient mechanical extractors for sustainable fibre retting

Sl. no.	Extraction method	Advantages	References
1	Gender-friendly power ribboner for extraction of green ribbon/bast from jute	Retting water saving by 50–60% and energy-efficient machine	Shambhu et al. [45]
2	CRIJAF Bast fibre extractor for jute and mesta	Retting water and energy saving in fibre production	Naik et al. [46]
3	Kenaf decorticator	Retting water and energy saving in fibre production	Ayorinde et al. [47]

Fig. 5 Key innovations in luxury textile treatment processes

printing provide precise, efficient, and customizable options for high-end textile production. These advancements not only enhance the functionality and appearance of textiles but also play a role in global sustainability initiatives. Figure 5 highlights key progress in luxury textile treatment processes.

3.3.1 Laser Technology

Laser technology, which stands for Light Amplification by Stimulated Emission of Radiation, has transformed the treatment of luxury textiles by providing a sustainable and efficient alternative to conventional methods. By using high-energy, focused light beams, laser systems can accurately alter fabric surfaces without the need for water or harmful chemicals, which aligns with the fashion industry's increasing focus on eco-friendly practices. This cutting-edge approach allows for the creation of intricate patterns and designs on a variety of materials, including natural fibres like cotton and synthetic textiles, thereby enhancing the visual appeal of luxury garments [48].

One of the main benefits of laser technology is its capacity to eliminate water usage, a significant concern in traditional textile finishing processes that often require large amounts of water for dyeing and finishing treatment. Laser treatments function in a dry environment, greatly reducing water consumption and lessening environmental impact. For example, laser engraving can produce detailed effects that would usually require multiple chemical baths, thus streamlining production while conserving resources [49, 50].

Energy efficiency is another important advantage. The non-contact nature of lasers allows for quick processing, which decreases overall energy consumption compared to traditional methods that involve heating water or using energy-intensive machinery. Laser systems provide high-quality results with lower energy use, assisting manufacturers in cutting costs and meeting sustainability goals by reducing their carbon footprint [51, 52].

Moreover, laser technology's versatility allows for multiple operations—such as cutting, engraving, and marking—within a single production cycle. This integration improves productivity, reduces waste, and minimizes resource use. For example, advanced systems like Jeanologia's Twin Super optimize production by simultaneously treating garments with dual laser tubes [53].

Beyond operational benefits, laser treatments enhance the quality of luxury textiles. They improve fabric properties such as drapability and wrinkle resistance without compromising material integrity. Moreover, the absence of harmful by-products contributes to a safer workplace for textile manufacturing employees [54].

3.3.2 ECOFAST Pure

ECOFAST Pure, developed by Dow Chemical, is a breakthrough in sustainable textile processing, particularly for luxury fabrics. This innovative technology addresses the environmental challenges of traditional dyeing, which often involves high water and energy consumption. By enhancing the absorption of dyes by natural fibres, ECOFAST Pure significantly reduces resource usage—cutting process chemicals by up to 90%, water by 50%, dye by 50%, and energy by 40%, all while maintaining exceptional colour quality and fastness [55–57].

The process involves pre-treating cotton fabrics with a cationic treatment that increases their affinity for negatively charged dyes. This enhances colour brightness, improves retention, and allows for dyeing at room temperature, requiring fewer rinses. These efficiencies conserve water, reduce energy demands, and achieve vibrant colours previously unattainable on cotton, making ECOFAST Pure a transformative innovation in the luxury textile sector where aesthetics and sustainability are critical [58, 59].

ECOFAST Pure also supports the fashion industry's sustainability goals by complying with the Zero Discharge of Hazardous Chemicals (ZDHC) Roadmap to *Zero Programme* and the Manufacturing Restricted Substances List. Dow's collaboration with Ralph Lauren to open-source a guide for using ECOFAST Pure encourages industry-wide adoption, promoting eco-friendly practices while maintaining the high standards expected in luxury textiles [54].

3.3.3 Plasma Technology

Plasma technology has emerged as a transformative solution for water- and energy-efficient textile colouration, particularly appealing to the luxury segment. Traditional dyeing methods are notorious for excessive water usage and reliance on harmful chemicals, resulting in significant environmental impacts. Plasma treatment offers a sustainable and cleaner alternative, aligning with the rising demand for eco-friendly practices in textile manufacturing.

The core principle of plasma technology involves using ionized gases (plasmas) to modify the surface properties of textile fibres. This surface modification enhances dye uptake, enabling superior dyeing performance for both natural and synthetic materials. Plasma treatment enhances the wettability of fibres, enabling dyes to penetrate more efficiently. Research shows that plasma-treated wool can achieve dye exhaustion rates that are nearly 50% higher than those of untreated wool. This improvement is attributed to the creation of reactive sites on the fibre surface, which

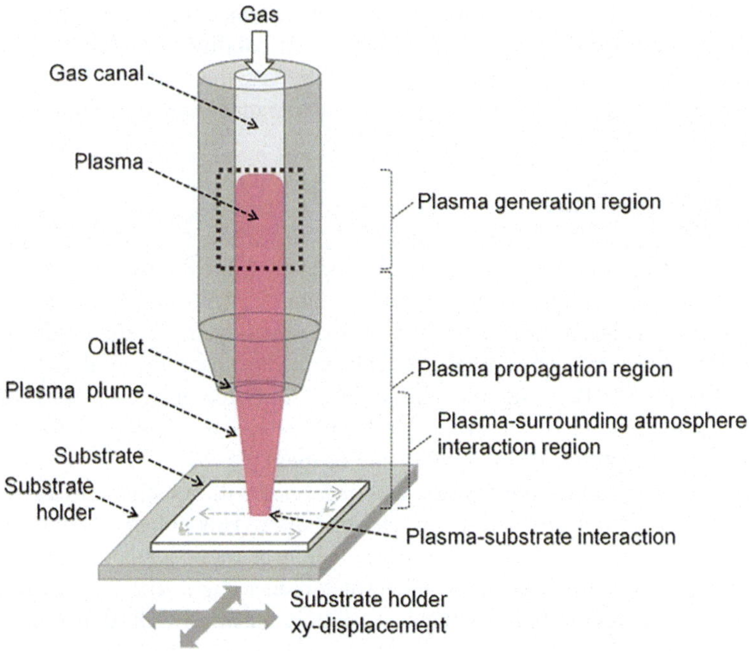

Fig. 6 A schematic representation illustrating the application of atmospheric plasma on textile materials. (Reproduced from Peran and Ercegović [62])

facilitate better dye bonding [60, 61]. Figure 6 illustrates a schematic representation of how atmospheric plasma is applied to textile materials, as shown in the study by Peran and Ercegović [62].

One of the key advantages of plasma technology is its efficient water and energy consumption. Unlike conventional dyeing, which requires extensive rinsing and high-temperature processes, plasma treatments typically operate under ambient conditions and are completed in seconds. This drastically reduces water usage and energy costs associated with heating large water volumes. In addition, the dry nature of plasma processes eliminates the need for chemical auxiliaries, thereby addressing wastewater pollution—one of the most critical environmental challenges in the textile industry [61, 63].

Plasma treatment is particularly suited for luxury textiles, where maintaining vibrant colours and fabric integrity is essential. It enhances colour fastness and surface properties without compromising the bulk characteristics of fabrics. For instance, cold plasma treatment of polyester fabrics has been shown to increase surface roughness, resulting in improved dye absorption and enhanced colour retention—qualities highly valued in high-end applications [63, 64].

Despite its potential, plasma technology faces challenges in adoption. High initial equipment costs can deter smaller manufacturers, and scaling atmospheric plasma systems for continuous, large-scale production remains a hurdle. However, as technological advancements drive cost reductions and improve scalability, plasma treatment is expected to play a larger role in sustainable textile manufacturing, particularly in the luxury market [61, 65].

3.3.4 Supercritical Carbon Dioxide (SC-CO_2) Dyeing of Textiles

SC-CO_2 dyeing is transforming the way luxury textiles are coloured, providing a sustainable and high-performance alternative to traditional dyeing techniques. This cutting-edge technology uses carbon dioxide in its supercritical state as a solvent, delivering rich, even colour while tackling the environmental issues typically linked to textile dyeing. The supercritical state occurs at temperatures above 31.1 °C and pressures over 73 atmospheres, CO_2 displays both gas-like and liquid-like characteristics. This distinctive combination enables it to easily penetrate fine textile fibres, and its high solvent capacity ensures effective dye dissolution and uniform distribution, making it ideal for luxury materials such as silk, cashmere, and other premium fibre blends [54, 66].

The dyeing process involves placing these valuable textiles in a sealed autoclave, where SC-CO_2 and the chosen dye are introduced. Acting as both a medium and a carrier, the SC-CO_2 promotes deep dye infusion into the fibres without the need for water or harmful chemicals. After the dyeing cycle is finished, depressurization converts the SC-CO_2 back to its gaseous form (CO_2), leaving behind beautifully dyed textiles without any wastewater or residual solvents. This method not only completely eliminates water usage but also operates at lower temperatures, helping to maintain the structural integrity and softness of delicate luxury fabrics. The outcome is a stunning, vibrant colouration with outstanding colour fastness to light, washing, and perspiration—qualities that are highly sought after in premium textiles [54, 67].

Innovative solutions from companies like DyeCoo Textile Systems highlight the potential of SC-CO_2 dyeing in the luxury market. Their advanced machines, such as the DyeOx4, allow for precise dyeing across a variety of fabric types, including knits, woven textiles, and yarns, all without using water or auxiliary chemicals. This method aligns with the increasing demand for eco-friendly luxury products, merging sustainability with outstanding aesthetic and tactile qualities [68].

Recent studies have further confirmed the benefits of SC-CO_2 dyeing for luxury applications. For example, research by Schmidt-Przewozna and Rój [69] showed that using SC-CO_2 dyeing with natural dyes, like madder, on linen and silk fabrics resulted in superior dye absorption and retention. These textiles demonstrated improved resistance to environmental factors such as sweat, washing, and light exposure while also requiring minimal energy and no water during the process. These findings highlight SC-CO_2 dyeing's ability to satisfy the dual demands of sustainability and high quality in the luxury textile industry.

By eliminating water use, cutting down on energy consumption, and providing exceptional dye quality, SC-CO_2 dyeing is set to become a key technology in the sustainable evolution of luxury textile production. This innovative method caters to the changing preferences of discerning consumers who appreciate both environmental responsibility and exquisite craftsmanship, establishing a new standard for eco-conscious elegance in the industry [69, 70].

3.3.5 Enzymatic Textile Process

Enzymes have transformed luxury textile processing by offering sustainable and efficient alternatives to chemical-intensive methods [71]. These biological catalysts enable premium-quality finishes while minimizing environmental impact through reduced water and energy use [72–75].

- **Cellulases**: Enhance softness, smoothness, and lustre by removing surface fuzz and pills, while also preventing pilling for improved durability.
- **Proteases**: Refine protein-based fibres like silk and wool by reducing felting, improving drape, and degumming silk for a softer, shinier finish. They also brighten colours by removing impurities.
- **Pectinases**: Improve absorbency and softness in plant-based fibres by removing natural impurities during bio-scouring.
- **Laccases**: Provide eco-friendly bleaching and selective colour modification for sustainable and creative design possibilities.
- **Amylases**: Efficiently desize fabrics, enhancing absorbency and preparing them for further processing.

Enzyme-based treatments reduce water and energy demands by enabling low-temperature processes with fewer rinse cycles. This aligns with the luxury textile industry's need for sustainable, high-quality production that meets the aesthetic and tactile standards of premium fabrics.

3.3.6 Digital Printing

Digital printing has transformed luxury textile production by significantly improving water and energy efficiency compared to traditional methods. Unlike screen printing, which consumes large amounts of water and energy, digital printing reduces water use by up to 95% and energy consumption by 75%. This efficiency is due to the precise application of ink directly onto fabric, minimizing waste [76–78].

A key advantage of digital printing is its use of water-based inks, which are less toxic and reduce volatile organic compound emissions, eliminating the need for harmful chemical solvents. Furthermore, digital printing eliminates energy-intensive pre-press processes, further enhancing sustainability [79].

The technology also allows for intricate designs with minimal ink waste—only 5–9 cc of ink is needed per meter of fabric, compared to 50–60 cc for screen printing. Digital printing also supports on-demand production, reducing waste from overproduction. As sustainability becomes a priority, digital printing is increasingly favoured in luxury textiles, offering eco-friendly solutions without compromising design quality [76, 77].

3.3.7 Foam Dyeing and Finishing for Luxury Apparel

Foam dyeing is an innovative, energy-efficient technology gaining grip in the textile industry, particularly for luxury apparel. Unlike conventional dyeing methods that rely heavily on water, foam dyeing replaces water with foam as the medium for applying dyes and finishing agents. This shift reduces water usage, minimizes waste, and enhances productivity, making it an environmentally sustainable alternative. The technology's low wet pickup significantly decreases the energy required for drying and allows for faster production cycles [80–82].

The principle of foam dyeing involves generating a stable foam containing dye liquor and applying it uniformly across the fabric. As the foam collapses, the dye liquor penetrates the textile substrate, facilitating effective dye fixation. Key parameters such as bubble size, foam stability, and viscosity are meticulously managed to ensure reliable outcomes. The process can be executed in either batch or continuous modes, with foam generation usually achieved through mechanical agitation or air injection [83]. For luxury fabrics like silk and premium cotton, foam dyeing provides precise control over dye distribution, guaranteeing even colouration without compromising the delicate texture of the material. The controlled application of foam minimizes mechanical stress on the fibres, helping to maintain their structural integrity and luxurious feel. Additionally, foam dyeing is compatible with various dye classes, including reactive and pigment dyes, which allows for versatility in design and colouration [84]. Figure 7 depicts the typical foam dyeing process for textile materials, followed by steam fixation, soaping, and washing. A significant benefit of foam dyeing is its substantial reduction in water and energy consumption. The wet pickup is considerably lower between 40% and 80% less than traditional methods. This not only shortens drying time but also decreases the carbon footprint

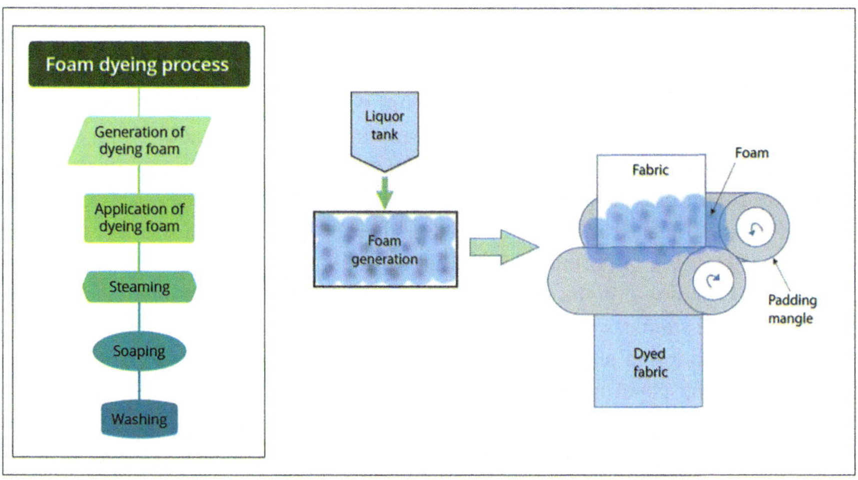

Fig. 7 Typical foam dyeing process. (Adapted from Abate and Tadesse [81])

of the process. Researches showed that foam dyeing can save up to 70–80% in energy and 30–90% in water usage, leading to considerable cost savings and a reduced environmental impact [84, 85].

Foam dyeing also facilitates advanced finishing processes. For instance, luxury clothing often needs specific finishes such as wrinkle resistance or softening agents, which can be effectively applied through foam. This technology guarantees an even application of these finishing agents while preserving the fabric's high quality. Foam dyeing has been successfully utilized in industries like carpet manufacturing and is now making its way into high-end fashion [81]. Additionally, foam dyeing requires minimal investment in new machinery since it can be integrated with existing equipment such as padders, stenters, and dryers. This adaptability, along with its environmental advantages, positions it as a practical option for luxury brands looking to boost sustainability without sacrificing quality or aesthetics [86–88].

3.3.8 Air Dyeing Technology for Textiles

Air dyeing technology has revolutionized the textile industry by providing a sustainable and efficient alternative to traditional dyeing methods, particularly for luxury textile treatment. This innovative process utilizes air as the primary medium for dye application, significantly reducing both water and energy consumption. According to various studies, air dyeing can reduce water usage by up to 95% and energy consumption by approximately 86% compared to conventional dyeing processes. Moreover, using this technology a single garment can save up to 170 L of water [89–91]. This remarkable efficiency not only conserves vital resources but also minimizes the environmental pollution typically associated with textile dyeing, which is responsible for a significant portion of industrial waste [88].

The operational principle of air dyeing involves atomizing dye liquor and mixing it with high-pressure airflow, which is then sprayed onto the fabric. The flowchart depicting the air dyeing process for synthetic textiles is presented in the Fig. 8 [92]. This method eliminates the need for a water bath, allowing for the direct application of dyes into the fibres, which enhances colour penetration and fixation. As a result, air-dyed fabrics exhibit superior colour fastness and vibrancy, addressing common issues such as fading and bleeding that often occur with conventional dyeing techniques. Furthermore, the process generates minimal fabric damage—only about 1% compared to 10% in traditional methods—thereby increasing yield and reducing waste [90, 91]. It is important to note that airflow or air-jet dyeing is distinct from air dyeing technology. In the airflow system, textile substrates are dyed in rope form within a closed stainless steel cylindrical vessel. Conversely, air dyeing technology utilizes sublimation, a process where inks are transferred to fabrics using specially designed transfer paper under heat and pressure [81, 88].

One of the key benefits of air dyeing is its versatility across different fabric types, including the delicate materials often found in luxury textiles. The technology not only supports environmentally friendly practices but also aligns with consumer demand for high-quality, durable fabrics that maintain their aesthetic appeal over

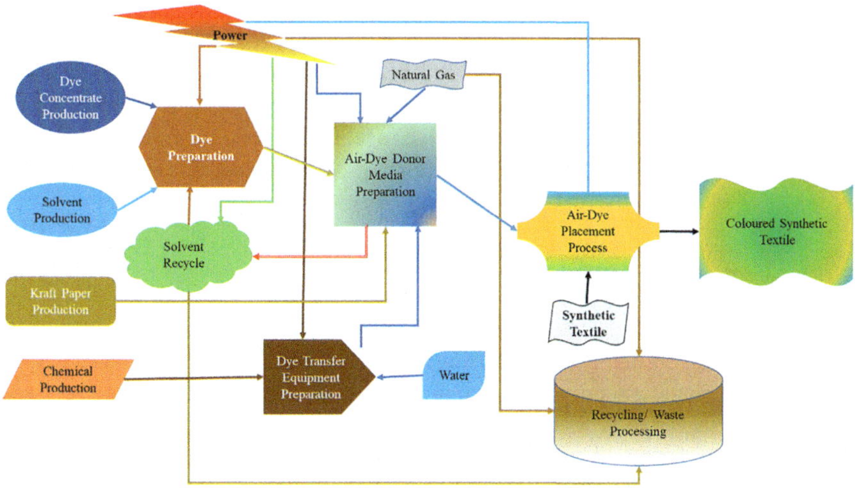

Fig. 8 Flow diagram for air-dye process of synthetic textile [92]

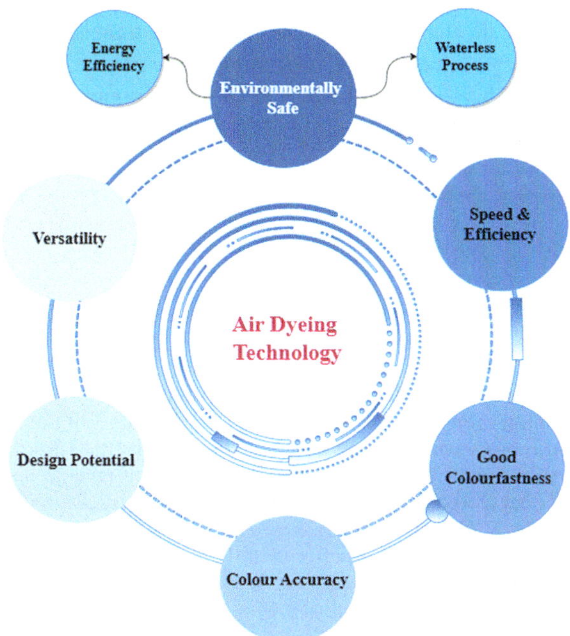

Fig. 9 Some of the features of air-dyeing technology [88]

time. Furthermore, air-dyed textiles can be washed with other garments without risk of colour transfer, as the dyes are embedded within the fibres rather than sitting on the surface [87, 90]. Figure 9 illustrates some features of air dye technology [88].

Despite its numerous benefits, challenges remain in the broader adoption of air dyeing technology. The initial investment in specialized equipment may deter some manufacturers; however, as sustainability becomes increasingly prioritized within the industry, more companies are likely to explore this innovative solution. Furthermore, ongoing advancements in dye formulations and machine designs continue to enhance the effectiveness and efficiency of air dyeing processes [88].

4 Conclusion

The manufacture of luxury textiles using water- and energy-efficient technologies is a critical step in the direction of a sustainable future for the textile sector. By implementing cutting-edge techniques including enzymatic procedures, green energy use, and novel treatment technologies, luxury brands are lessening their environmental impact without sacrificing their dedication to exceptional quality and craftsmanship. In addition to conserving essential resources, techniques like digital printing, SC-CO_2 dyeing, and sustainable retting procedures also establish new standards for efficiency and innovation. These developments provide eco-friendly solutions that satisfy the demands of discriminating customers, reaffirming that sustainability and luxury can coexist. An environmentally conscious and ethical luxury textile business will be shaped in large part by the ongoing development of these technologies.

References

1. Leal Filho, W., Perry, P., Heim, H., Dinis, M. A. P., Moda, H., Ebhuoma, E., & Paço, A. (2022). An overview of the contribution of the textiles sector to climate change. *Frontiers in Environmental Science, 10*, 973102. https://doi.org/10.3389/fenvs.2022.973102
2. Uddin, M. A., Begum, M. S., Ashraf, M., Azad, A. K., Adhikary, A. C., & Hossain, M. S. (2023). Water and chemical consumption in the textile processing industry of Bangladesh. *PLOS Sustainability and Transformation, 2*(7), e0000072. https://doi.org/10.1371/journal.pstr.0000072
3. Hasanbeigi, A., & Price, L. (2015). A technical review of emerging technologies for energy and water efficiency and pollution reduction in the textile industry. *Journal of Cleaner Production, 95*, 30–44. https://doi.org/10.1016/j.jclepro.2015.02.079
4. Saif, M., Blay-Roger, R., Zeeshan, M., Bobadilla, L. F., Reina, T. R., Nawaz, M. A., & Odriozola, J. A. (2024). Navigating the legislative interventions, challenges, and opportunities in revolutionizing textile upcycling/recycling processes for a circular economy. *ACS Sustainable Resource Management, 1*(11), 2338–2349.
5. Schönberger, H., & Schäfer, T. (2003). *Best available techniques in textile industry*. Federal Environmental Agency. https://citeseerx.ist.psu.edu/document?repid=rep1&type=pdf&doi=a7c4ddb9f9dad001c4e9a70fcd49ed2636425262
6. Desore, A., & Narula, S. A. (2018). An overview on corporate response towards sustainability issues in textile industry. *Environment, Development and Sustainability, 20*, 1439–1459.

7. Roy, M., Sen, P., & Pal, P. (2020). An integrated green management model to improve environmental performance of textile industry towards sustainability. *Journal of Cleaner Production, 271*, 122656.
8. European Commission. (2003). *Energy consumption in the service sector: Survey of EU member states*. https://ec.europa.eu/eurostat/documents/3217494/5634673/KS-43-02-042-EN.PDF/32e10ad4-43a3-4635-97f3-01b28ed6ba18
9. Dalle, Juhriyansyah/Hayat, Atma et al. (2021). The influence of accounting information system and energy consumption on carbon emission in the textile industry of Indonesia: Mediating role of the supply chain process. International Journal of Energy Economics and Policy 11(1):536–543
10. Hasanbeigi, A., & Price, L. (2012). A review of energy use and energy efficiency technologies for the textile industry. *Renewable and Sustainable Energy Reviews, 16*(6), 3648–3665. https://doi.org/10.1016/j.rser.2012.03.029
11. Çay, A. (2018). Energy consumption and energy saving potential in clothing industry. *Energy, 159*, 74–85. https://doi.org/10.1016/j.energy.2018.06.128
12. Buckley, S. (2022). *The textile Industry's transition to renewable energy*. https://www.fibre2fashion.com/industry-article/9496/the-textile-industry-s-transition-to-renewable-energy
13. Gillow, J., Sentance, B., & Sentance, B. (1999). *World textiles: A visual guide to traditional techniques*. Thames & Hudson. https://pdfs.semanticscholar.org/2967/a74339f4bfb51c28403 5eb7780feb8563db9.pdf
14. Jayawardena, N. S., Quach, S., Perera, C. H., Thaichon, P., & Dewasiri, N. J. (2023). Luxury marketing and sustainability in the south Asian context. In *Luxury marketing, sustainability and technology* (pp. 1–14). Routledge. https://pure.jgu.edu.in/id/eprint/5828/1/9781003321378_previewpdf.pdf
15. Kim, D. D. I. (2024). *Beyond natural dyes: Embracing sustainable innovations in industrial textile dyeing*. https://doi.org/10.26434/chemrxiv-2024-ztldb-v2
16. Gunasekar, V., & Ponnusami, V. (2015). Eco-friendly textile dyeing processes. In *Hydrogen production and remediation of carbon and pollutants* (pp. 255–287). Springer. https://doi.org/10.1007/978-3-319-19375-5_6
17. Verma, H., Sharma, A. K., Rani, A., Thiyam, G., & Pavan, M. (2023). Optimization of the process parameters for dyeing silk fabric with a fungal pigment using RSM. *Sustainable Chemistry and Pharmacy*, 35, 101192. https://doi.org/10.1016/j.scp.2023.101192
18. Vilumsone-Nemes, I., Kaplan, V., & Belakova, D. (2020). Potentialities of reducing textile waste in the manufacturing of garments from striped fabrics. *Fibres & Textiles in Eastern Europe, 28*(6), 58–63. https://doi.org/10.5604/01.3001.0014.3799
19. Kaur, M., Pavan, M., & Samant, L. (2024). Eco Textiles: The Present and the Future. In *Climate Action Through Eco-Friendly Textiles* (pp. 179–206). Singapore: Springer Nature Singapore. https://doi.org/10.1007/978-981-99-9856-2_12
20. Farhana, K., Kadirgama, K., Mahamude, A. S. F., & Mica, M. T. (2022). Energy consumption, environmental impact, and implementation of renewable energy resources in global textile industries: An overview towards circularity and sustainability. *Materials Circular Economy, 4*(1), 15. https://doi.org/10.1007/s42824-022-00059-1
21. Bose, S., Bharaskar, S., Khubalkar, S., & Pahariya, Y. (2023). Energy audit of textile industry. In *Recent advances in material, manufacturing, and machine learning* (pp. 1321–1327). CRC Press.
22. Paulose, T. (2024). Enhancing textile production with energy-efficient modern compressors. *Indian Textile Journal*. https://indiantextilejournal.com/enhancing-textile-production-with-energy-efficient-modern-compressors/
23. Sharma, S. (2012). Energy management in textile industry. *International Journal of Power System Operation Energy Management ISSN (PRINT), 2*(1), 2.
24. Elwardany, M. (2024). Enhancing steam boiler efficiency through comprehensive energy and exergy analysis: A review. *Process Safety and Environmental Protection, 184*, 1222–1250. https://doi.org/10.1016/j.psep.2024.01.102

25. Owodiong-Idemeko, B. M., Adubi, K. O., & Owodiong, U. (2022). Analysis of the benefits and challenges faced in skill acquisition in the utilization of green energy in the clothing and textiles laboratory in tertiary institutions in Lagos state. *Journal of Association of Vocational and Technical Educators of Nigeria, 28*(1).
26. Islam, M. A. (2021). *Energy auditing and efficiency of a ready–made garment factory in Bangladesh: A case study* (Doctoral dissertation, Brac University). http://hdl.handle.net/10361/15657
27. Hoque, M. T., Mazumder, N. U. S., & Islam, M. T. (2021). Enzymatic wet processing. In *Sustainable practices in the textile industry* (pp. 87–110). Wiley. https://doi.org/10.1002/9781119818915.ch4
28. Hasan, M. M., Nabi, F., & Mahmud, R. (2015). Benefits of enzymatic process in textile wet processing. *International Journal of Fiber and Textile Research, 5*(2), 16–19.
29. Mojsov, K. (2012). Enzyme scouring of cotton fabrics: A review. *International Journals of Marketing and Technology, 2*(9), 256–275.
30. Chen, J., Wang, Q., Hua, Z., & Du, G. (2006). Research and application of biotechnology in textile industries in China. *Enzyme and Microbial Technology, 40*(7), 1651–1655. www.asean-biotechnology.info/Abstract/21026447.pdf
31. Jajpura, L., & Nayak, R. (2020). Ultrasound applications in textiles and apparels. In *Sustainable technologies for fashion and textiles* (pp. 143–161). Woodhead Publishing. https://doi.org/10.1016/B978-0-08-102867-4.00007-4
32. Ramachandran, T., Karthik, T., & Saravanan, D. (2008). Novel trends in textile wet processing. *Journal of Institution of Engineers, 89*, 3.
33. Harifi, T., & Montazer, M. (2015). A review on textile sonoprocessing: A special focus on sonosynthesis of nanomaterials on textile substrates. *Ultrasonics Sonochemistry, 23*, 1–10. https://doi.org/10.1016/j.ultsonch.2014.08.022
34. Benli, H., & Bahtiyari, M. I. (2015). Combination of ozone and ultrasound in pretreatment of cotton fabrics prior to natural dyeing. *Journal of Cleaner Production, 89*, 116–124. https://doi.org/10.1016/j.jclepro.2014.11.007
35. Eren, H. A., & Ozturk, D. (2011). The evaluation of ozonation as an environmentally friendly alternative for cotton preparation. *Textile Research Journal, 81*(5), 512–519. https://doi.org/10.1177/0040517510380782
36. Banchero, M. (2020). Recent advances in supercritical fluid dyeing. *Coloration Technology, 136*(4), 317–335. https://doi.org/10.1111/cote.12469
37. Long, J. J., Xu, H. M., Cui, C. L., Wei, X. C., Chen, F., & Cheng, A. K. (2014). A novel plant for fabric rope dyeing in supercritical carbon dioxide and its cleaner production. *Journal of Cleaner Production, 65*, 574–582. https://doi.org/10.1016/j.jclepro.2013.08.008
38. Jun, J. H., Sawada, K., & Ueda, M. (2004). Application of perfluoropolyether reverse micelles in supercritical CO_2 to dyeing process. *Dyes and Pigments, 61*, 17–22. https://doi.org/10.1016/j.dyepig.2003.08.003
39. Hussain, T., & Wahab, A. (2018). A critical review of the current water conservation practices in textile wet processing. *Journal of Cleaner Production, 198*, 806–819. https://doi.org/10.1016/j.jclepro.2018.07
40. Muhib, M. I., & Rahman, M. M. (2024). Water and energy efficiency in textile production. In P. Singh (Ed.), *Dye pollution from textile industry. SDGs and textiles*. Springer. https://doi.org/10.1007/978-981-97-5341-3_14
41. Shambhu, V. B., Thakur, A. K., Nayak, L., & Das, B. (2018). NIRJAFT power Ribboner for improved retting of jute and mesta plants. In *Proceedings of the National Seminar on market driven innovations in natural Fibers, ICAR-National Institute of research on jute and allied Fiber technology*, Kolkata, India (Vol. 29)
42. Majumdar, B., Barai, S., Paswan, A., Naik, R. K., Mazumdar, S. P., Sarkar, S., et al. (2024). Jute retting using free-flowing water with microbial consortium minimizes water requirement and water pollution. *Water Conservation Science and Engineering, 9*(2), 36. https://doi.org/10.1007/s41101-024-00270-y

43. Rana, M. R. K., Amin, A., Ayub, A., Shaon, F. H., Paul, J. C., Chowdhury, F. T., et al. (2024). An improved jute ribbon retting using microbial consortia from jute retting water. *Industrial Crops and Products, 221*, 119335. https://doi.org/10.1016/j.indcrop.2024.119335
44. Ray, D. P., Saha, S. C., Sarkar, A., & Ghosh, R. K. (2016). Production of quality jute fibre through accelerated retting. *International Journal of Bioresource Science, 3*(2), 57–65. https://doi.org/10.5958/2454-9541.2016.00016.5
45. Shambhu, V. B., Shrivastava, P., Jagadale, M., Nayak, L. K., & Shakyawar, D. B. (2023). Development of gender-friendly power ribboner for extraction of green ribbon/bast from jute plants. *Journal of Natural Fibers, 20*(2), 2250076. https://doi.org/10.1080/15440478.2023.2250076
46. Naik, R. K., Mazumdar, S. P., Majumdar, B., & Behera, M. S. (2018). Mechanical extraction of jute and mesta for quality fibre production. *Journal of the Indian Society of Coastal Agricultural Research, 36*(1), 76–79.
47. Ayorinde, A. T., Makanjuola, G. A., Aluko, O. B., Owolarafe, O. K., & Sanni, L. A. (2019). Performance evaluation of a kenaf decorticator. *Agricultural Engineering International: CIGR Journal, 21*(1), 192–202.
48. Unal, F., Yavas, A., & Avinc, O. (2020). Sustainability in textile design with laser technology. In *Sustainability in the textile and apparel industries: Sustainable textiles, clothing design and repurposing* (pp. 263–287). Springer. https://doi.org/10.1007/978-3-030-37929-2_11
49. Atwee, T., Elshafei, N., Mostafa, H., & Bassyouni, F. (2022). Application of laser technology to the surface treatments of various synthetic textile samples. *World Journal of Advances in Applied Physics and Mathematical Theories, 1*, 1–11. https://scientificeminencegroup.com/articles/JAPM/Application-of-Laser-Technology-to-the-Surface.pdf
50. Davis, R. (2020, July 23). *Rachael Davis*. https://www.textileworld.com/textile-world/features/2020/07/laser-finishing-for-textile-materials/
51. Scantech Laser. (2024, October 4). *Textile & Leather*. https://scantechlaser.com/solutions/textile-leather-industry/
52. Jeanologia. (2024). *Laser marking technologies for textile industry*. https://www.jeanologia.com/garment/laser/
53. Poddar, A. (2021, November 26). *How laser technology is a boon for the textile industry?* Cheer Sagar. https://www.cheersagar.com/blog/how-laser-technology-is-a-boon-for-the-textile-industry
54. Sakib, N. (2023, July 19). *Reducing carbon footprints and water consumption through sustainable dyeing and finishing*. TexSPACE Today. https://www.texspacetoday.com/reducing-carbon-footprints-water-consumption-sustainable-dyeing-finishing/
55. *Dow chemical introduces sustainable textile processing: EcoFast Pure reduces water and energy consumption during dyeing*. (2018). http://vitajeans.com/news/industry-news/dow-chemical-introduces-sustainable-textile-processing-ecofast-pure-reduces-water-and-energy-consumption-during-dyeing-143.html
56. Lauren, R., & Susko, K. (2021). *Ralph Lauren and Dow open-source new process to transform how the fashion industry dyes cotton*. https://corporate.ralphlauren.com/pr_211013_RalphLaurenxDow.html
57. Dow Corporate. (n.d.). *ECOFASTTM*. https://www.dow.com/en-us/brand/ecofast.html
58. Editorial Team. (2018, September 7). *ECOFAST textile treatment technology by Dow Chemical*. Apparel Views. https://www.apparelviews.com/ecofast-textile-treatment-technology-by-dow-chemical/
59. Gardner, F., & Dranttel, J. (2021). *Dow ECOFASTTM pure: Sustainable cotton pretreatment* | Knowledge Hub | Circle Economy Foundation. https://knowledge-hub.circle-economy.com/article/9410?n=Dow-ECOFAST%E2%84%A2-Pure-Sustainable-Cotton-Pretreatment
60. Jelil, R. A. (2015). A review of low-temperature plasma treatment of textile materials. *Journal of Materials Science, 50*(18), 5913–5943. https://doi.org/10.1007/s10853-015-9152-4
61. Narsi. (2024). *Low-emission textile dyeing and finishing*. Climate Innovation Series. https://www.climafix.in/ref/cis/innovation/low-emission-textile-dyeing-and-finishing/

62. Peran, J., & Ercegović Ražić, S. (2020). Application of atmospheric pressure plasma technology for textile surface modification. *Textile Research Journal, 90*(9–10), 1174–1197. https://doi.org/10.1177/0040517519883954
63. Vibe Fabriclore. (2023, January 28). *Sustainable dyeing and finishing methods in textile industry*. Fabriclore. https://fabriclore.com/blogs/textiles/sustainable-dyeing-and-finishing-methods-in-textile-industry
64. Sphere, T., & Prajapati, G. (2021, February 26). *Plasma technology in textiles: Textile applications*. Textile Sphere. https://www.textilesphere.com/2019/08/plasma-technolgy-in-textiles.html
65. Naveed, M. (2018, April 3). *Plasma treatment as green technology for dyeing of textile fabrics*. Crimson Publishers. https://crimsonpublishers.com/tteft/fulltext/TTEFT.000534.php
66. Sami, S. K. (2023, July 25). *Waterless carbon dioxide dyeing is a powerful solution for sustainable coloration*. TexSPACE Today. https://www.texspacetoday.com/waterless-carbon-dioxide-dyeing/
67. Kannan, S. S. (2007). *Super critical carbon dioxide assisted dyeing—A novel approach*. https://www.fibre2fashion.com/industry-article/1678/super-critical-carbon-dioxide-assisted-dyeing
68. Dyecoo. (2022, March 30). *CO_2 dyeing*. https://dyecoo.com/co2-dyeing/
69. Schmidt-Przewozna, K., & Rój, E. (2023). Green sustainable textile supercritical dyeing process using CO_2 madder (Rubia t inctorum L.) extract. *Journal of Natural Fibres, 20*(2), 2277836. https://doi.org/10.1080/15440478.2023.2277836
70. Kiron, M. I. (2022, September 5). *Supercritical fluid dyeing and its advantages*. Textile Learner. https://textilelearner.net/supercritical-fluid-dyeing-and-its-advantages/
71. Pavan, M., Samant, L., Mahajan, S., & Kaur, M. (2024). Role of chemicals in textile processing and its alternatives. In Climate action through eco-friendly textiles (pp. 55–72). Singapore: Springer Nature Singapore. https://doi.org/10.1007/978-981-99-9856-2_5
72. Samanta, K. K., Basak, S., & Chattopadhyay, S. K. (2015). Specialty chemical finishes for sustainable luxurious textiles. *Handbook of Sustainable Luxury Textiles and Fashion, 1*, 145–184. https://doi.org/10.1007/978-981-287-633-1_7
73. Cavaco-Paulo, A., & Gubitz, G. (Eds.). (2003). *Textile processing with enzymes* (Vol. 29). Woodhead Publishing.
74. Roy Choudhury, A. K. (2014). Sustainable textile wet processing: Applications of enzymes. In *Roadmap to sustainable textiles and clothing: Eco-friendly raw materials, technologies, and processing methods* (pp. 203–238). Springer. https://doi.org/10.1007/978-981-287-065-0_7
75. Riva, S. (2006). Laccases: Blue enzymes for green chemistry. *Trends in Biotechnology, 24*(5), 219–226. https://doi.org/10.1016/j.tibtech.2006.03.006
76. FESPA. (n.d.). *The environmental benefits of waterless digital textile print ink systems*. https://www.fespa.com/en/news-media/the-environmental-benefits-of-waterless-digital-textile-print-ink-systems
77. York, W. N. (2021). *The environmental benefits of digital printing*. https://www.wilengroup.com/news/the-environmental-benefits-of-digital-printing
78. Tanim, M. A. M. (2024, June 26). *Sustainable innovations in textile printing: A deep dive into water-based and eco-friendly inks*. TexSPACE Today. https://www.texspacetoday.com/water-based-and-eco-friendly-inks/
79. Konica Minolta. (2024, March 5). *How sustainable is digital printing really?* https://www.konicaminolta.eu/eu-en/ipp-blog/sustainability/how-sustainable-is-digital-printing-really
80. Gregorian, R. S., Bafford, R. A., & Namboodri, C. G. (1979). The utilization of foams in the wet processing of textiles. In *ACS symposium series* (pp. 155–173). ACS Publication. https://doi.org/10.1021/bk-1979-0107.ch013
81. Abate, M. T., & Tadesse, M. G. (2021). Airflow, foam, and supercritical carbon dioxide dyeing technologies. In *Innovative and emerging technologies for textile dyeing and finishing* (pp. 137–164). Wiley. https://doi.org/10.1002/9781119710288.ch5
82. Van Der Walt, G. H. J., & Van Rensburg, N. J. J. (1986). Low-liquor dyeing and finishing. *Textile Progress, 14*(2), 1–50. https://doi.org/10.1080/00405168608688900

83. Elbadawi, A. M., & Pearson, J. S. (2003). Foam technology in textile finishing. *Textile Progress, 33*(4), 1–31. https://doi.org/10.1080/00405160308688960
84. Yu, H., Wang, Y., Zhong, Y., Mao, Z., & Tan, S. (2014). Foam properties and application in dyeing cotton fabrics with reactive dyes. *Coloration Technology, 130*(4), 266–272. https://doi.org/10.1111/cote.12088
85. Chequer, F. M. D., De Oliveira, G. A. R., Ferraz, E. R. A., Carvalho, J., Zanoni, M. V. B., & De Oliveir, D. P. (2013). *Textile dyes: Dyeing process and environmental impact*. IntechOpen. https://doi.org/10.5772/53659
86. Chen, S., Wang, C., Fei, L., & Liu, H. (2017). A novel strategy for realising environmentally friendly pigment foam dyeing using polyoxyethylene ether surfactant C14 EO 5 as a foam controller. *Colouration Technology, 133*(3), 253–261. https://doi.org/10.1111/cote.12272
87. Mahmud, I., & Kaiser, S. (2020). Recent progress in waterless textile dyeing. *Journal of Textile Science & Engineering, 10*(6). https://doi.org/10.37421/jtese.2020.10.421
88. Khalil, E., Sarkar, J., Rahman, M. M., Shamsuzzaman, M., & Das, D. (2023). Advanced technology in textile dyeing. In *Advanced technology in textiles: Fibre to apparel* (pp. 97–138). Springer Nature. https://doi.org/10.1007/978-981-99-2142-3_4
89. DEBS Corporation. (n.d.). *Textile and lifestyle solutions*. https://debscorp.com/textiles/airdye/
90. Palamutcu, S. (2017). Sustainable textile technologies. In *Textiles and clothing sustainability: Sustainable technologies* (pp. 1–22). Springer.
91. Nayak, R., Nguyen, L. V. T., Panwar, T., & Jajpura, L. (2020). Sustainable technologies and processes adapted by fashion brands. In *Sustainable technologies for fashion and textiles* (pp. 233–248). Woodhead Publishing.
92. Dhanabalan, V., Sukanya, & Lokesh, K. V. (2015, October 14). *Air-dyeing technology—A review*. Textile News, Apparel News, RMG News, Fashion Trends. https://www.textiletoday.com.bd/air-dyeing-technology-a-review

Summing Up: Luxury in Harmony with Sustainability

Sadhna, E Mokanaasri, Rajesh Kumar, and S. Greeshma

1 Introduction

Sustainability in luxury fashion has emerged as a critical theme making the concept of sustainable luxury become apparent as a focal area within the luxury sector. For a long time, luxury brands were symbols of exclusivity and lavishness, and while those notions still hold, there is a new expectation and desire for more from consumers regardless of Bachelor of Arts in Marketing Consumer Behavior. Millennials and Gen Z [1] consumers are more concerned with the values of the products that they purchase. This is one of the reasons why a luxury brand cannot solely depend on its exclusivity or luxuriousness to sustain rivalry in the current market. Data show that, by 2025, these demographics will account for about 70% of the luxury consumer base, meaning that luxury brands must incorporate green initiatives into strategies as central parts of their strategies for engaging these consumers [2].

A comparative overview of attention to sustainability in luxury analysis reveals that it is no longer an option to be merely aligned or a marketing ploy; it has become a core of focus. As a result, there is consciousness among luxury brands, and they nervously regulate sustainable practices such as using environment-friendly fabric, re-considering the packaging, and ethical labor practices. For example, in its current operations, such brands as the Patagonia brand have developed concrete strategies on how sustainability could be integrated into organizational culture through encouraging product repair instead of product replacement. In addition, newer strategies such as introducing blockchain technology to supply chain transparency can be seen to help consumers track the complete life cycle of a product, backing up brand promises [3].

Sadhna · E. Mokanaasri (✉) · R. Kumar · S. Greeshma
Department of Fashion Design, School of Arts and Design, Woxsen University, Hyderabad, Telangana, India

Sadhna et al. (eds.), *Crafting Sustainability in Luxury Textiles for a Zero-Waste Future*, Sustainable Textiles: Production, Processing, Manufacturing & Chemistry, https://doi.org/10.1007/978-3-031-86773-6_14

The need for such change is well timed given the increasingly prevalent consciousness of the negative effects of the fashion industry on the environment. The sector has come under the spotlight, particularly in relation to its impact on wastage and pollution. Some of the traditional practices in the production line have therefore required consideration. The idea of eco-luxury emerged as a response, where the luxury of the respective brands is achieved with excellent workmanship and the burndown of resources harming the earth. This blending of luxury with sustainability is not a fashion; it is a paradigm shift: a new layer of understanding of what it is to be luxurious in the present modern society [4].

While the affairs of the luxury textile industry are still in their social, political, and environmental infancy, there is a desperate need for zero-waste projects. The urgency of waste reduction strategies is explained by the past behavior of the industry, which has been characterized by *ad hoc* approaches aimed at achieving quick gains. The use of the concept of zero-waste for products demands eco-innovative resource strategies regarding the design of products and their production that reduce the generation of waste at the usage stage beginning from the source [5]. Future measures for the integration of waste elimination plans for the luxury industry will be a key concern of this chapter. Designers, manufacturers, and consumers are aware that systems with circularity of fashion items should be developed through everyone's collective effort. This can involve decisions as to which to reuse items and recycle materials and which to design products or packaging for durability and not obsolesce. Therefore, cooperation with the stakeholders is imperative to replace the luxury textile industry under the current model and create an industry that is both sustainable environmentally and economically healthy [6].

2 The Sustainability Challenge in Luxury Fashion

Luxury fashion is one of the most significant contributors to global climate crisis waste, producing about 10% of carbon emissions. Important environmental issues include resource depletion; the creation of high-quality luxury goods often encompasses natural resources: cotton production uses water in significant quantities, while synthetic fibers make microplastics. According to the Ellen MacArthur Foundation, one major issue is that 87% of garment fibers are found in landfills. In contrast, only 1% of clothing is recycled for reuse, meaning that the textile industry has to produce 92 million tons of annual waste. Other processes, such as dyeing and finishing, add more to the water pollution problem as it accounts for approximately 20% of global wastewater. In addition, because of ethics in using materials such as leather and fur, many luxury brands are now coming up with bio-textiles in order to decrease ecological and ethical impact [7, 8].

Consumer awareness of issues on sustainability has driven more consumers to make responsible choices that align with ethical and environmentally friendly products. According to a 2022 survey, 57% of luxury buyers now consider the sustainability of a purchase. It follows the emerging market trend that is increasingly considering serious commitment on the part of the brands to ethical and environmental concerns [9]. Added pressures of transparency and regulatory compliance accelerate this shift, such as the European Union's proposed Corporate Sustainability Due Diligence Directive requiring a brand to address specific negative impacts in its supply chains or face fines [10]. Luxury companies are embracing corporate responsibility and innovation by having sustainable practices in their firm, such as using recycled or bio-based materials, embracing the circular economy principle, or maybe making sure that supply chains are transparent; for instance, Kering aims to reduce greenhouse gas emissions by 40% by 2035, while Gucci has introduced a "circular hub" for products made from recycled materials. Another evidence of the economic feasibility of sustainability is that it creates a market for green products, thus making brands more competitive and creating a good public image [11].

3 Advancing Toward a Zero-Waste Future in Luxury Textile Crafting

Many luxury brands are incorporating sustainability through new product lines featuring recycled materials. For instance, Prada, which released the Re-Nylon collection in 2019, has developed a product that makes use of Econyl, regenerated nylon that originated from plastics collected from oceans and fishery plastics, aiming to eventually switch over all its products to Re-Nylon by late 2021, potentially avoiding 50 tons of virgin nylon each year [12]. Off The Grid is another line from Gucci, made with Econyl recycled nylon and recycled polyester thread in a closed-loop manufacturing process that repurposes production waste. Burberry launched a 2020 collection of tops in recycled cotton and trousers in polyester from plastic bottles, targeting to cut greenhouse gas emissions by 43% since 2016 and ensure 68% of its products meet sustainability standards. Louis Vuitton's Felt Line by Virgil Abloh is one of the designs that made the entire line using recycled polyester from plastics and pre-consumer waste [13].

Luxury brands are increasingly partnering with startups to create innovative, sustainable textiles, further advancing their commitment to green materials. The partnerships between luxury brands and innovators pool together the expertise of both parties to create durable, high-quality, green materials suited for the luxury market. For example, Stella McCartney has partnered with Bolt Threads, a biotechnology company, to produce Mylo, a substitute for leather made from the root system of mushrooms [12]. This mycelium replaces the traditional leather. Hermès collaborates with MycoWorks to create a Victoria shopper bag by using Sylvania, a leather alternative based on mycelium, tanned and finished in Hermès French tanneries to preserve quality. Such a company includes the luxury group Kering, owner of houses such as Gucci and Balenciaga, investing in sustainable textile startups, for instance, Renewcell, which produces biodegradable fibers from recycled cotton, and Kintra, which is developing plant-based compostable fibers [14].

3.1 Optimizing Production Processes

Zero-waste pattern-cutting techniques are a very important step forward for clothing production, aiming to cut down fabric waste at the cutting stage and designing patterns that have zero over leftover fabric to optimize the use of material. Key strategies include pattern design, in which Computer-Aided Design (CAD) technology is used to arrange pieces as a jigsaw puzzle for maximum fabric efficiency; modular cutting, in which pieces can be cut in multiples simultaneously to reduce waste; and fabric utilization analysis, in which consumption reports are gathered and analyzed for areas that waste, thus optimizing cutting arrangements concerning the characteristics of fabrics [15].

The primary focus of lean manufacturing processes is on enhancing productivity by removing waste from the supplier's involvement throughout the production value stream to minimize waste within the supply chain. Important principles include value stream mapping, where all the production activities are defined and documented to identify and correct waste and non-value-added tasks to allow targeted improvements in operations. Just-in-time inventory incorporates suppliers into a system where materials are used only as needed, decreasing waste stock and the costs of storage. Continuous improvement, or Kaizen, creates an environment where methods are continuously evaluated and refined, thereby promoting collaboration between producers and suppliers to achieve sustainable efficiency [16].

3.2 Future Directions in Luxury Textile Crafting

Innovative design in zero-waste is revolutionizing luxury textiles to focus on zero material waste, using groundbreaking methods. Modular clothing allows consumers to wear clothes in different styles, and parts may be interchanged without alteration for responsible resource utilization and waste prevention [17]. Upcycling is gaining popularity among luxury brands as the process of upcycling turns materials or products into unique, standalone items that catch the attention of environmentally conscious consumers while eliminating waste. Also, 3D printing technology revolutionizes fashion by allowing designers to use precise amounts of material for intricate structures, reducing material consumption compared to traditional methods and facilitating the creation of unique, desirable products with minimal waste [18].

3.3 Circular Economy in Luxury

Luxury fashion brands embrace circular economy practices to reduce waste and fight product obsolescence as consumers increasingly prefer sustainable purchasing. Such practices help enhance the brand image and encourage environmental stewardship. These include designing for longevity, whereby products are designed to last longer, thereby building long-term customer loyalty. Brands such as Gucci and Burberry are introducing resale and rental programs, encouraging customers to return used items for resale, reducing waste, and appealing to sustainability-conscious consumers. Closed-loop production supports circularity further by reusing materials from old products in new production, significantly reducing reliance on virgin resources and alleviating environmental strain [19].

The future of materials in the textile industry will shift toward biodegradable fabrics and regenerative fibers, revolutionizing luxury fashion by blending environmental responsibility with unwavering quality. Luxury brands increasingly feature these innovations, delivering a message of commitment to sustainability. Lab-grown textiles created using biotechnology are produced in laboratories rather than derived from fossil fuels, with minimum water usage and chemical waste [20]. Advances in eco-friendly dyes also further enhance sustainability by reducing water usage and eliminating hazardous substances in the dyeing process. In addition, bio-based fibers, sourced from renewable resources, are designed to biodegrade at the end of their lifecycle, reducing landfill waste and contributing to sustainable fashion practices.

3.4 Technology-Driven Sustainability

Technological innovation is changing the luxury textile industry, making it more sustainable, efficient, and accountable while enabling brands to meet consumers' increasingly demanding ethical standards. Blockchain technology is being integrated to ensure transparency in the supply chain, allowing customers to trace the provenance of raw materials and verify ethical production practices, thereby fostering trust and promoting responsible sourcing. Meanwhile, AI is transforming production management by cutting unnecessary expenses and resource wastage, allowing brands to focus on high-demand products, thereby supporting more sustainable practices [21].

4 Shaping a Sustainable Landscape for Luxury Fashion

The Fashion Pact, launched at the G7 summit in 2019, is a groundbreaking initiative that brings together over 160 brands with a shared commitment to achieving net-zero emissions by 2050. Figure 1 shows the Fashion pact focuses on three key pillars: the conservation of the atmosphere, living organisms, and water, aiming to mitigate the negative environmental impacts of the fashion industry. Many participating brands, such as Kering and Louis Vuitton Moët Hennessy (LVMH), have set up Science Based Targets initiative (SBTi), which works toward sustainable energy solutions. Implemented practices for operational sustainability management become

Fig. 1 Fashion Pact with three pillars [22]

essential for focusing on responsible sourcing to decrease the environmental footprint [23].

The arrangement also helps The Fashion Pact's member companies to learn from one another and encourages companies to report on sustainable practices. Because the Pact collects information about achievements and issues companies encounter, the brands can obtain information about the experience of their b-trade partners. This shared approach not only advances sustainability at a faster rate but also provides the right roadmap toward improvement across Luxury fashion.

This is why it is important for luxury fashion goods consumers to be engaged in order to create a sustainable culture among them. As the consciousness of consumers increases day by day about the problems affecting the environment, luxury brands have the role of making people aware of how fashion is affecting our planet. Out of the two, appropriate communication strategies that publicize sustainable practices should improve consumer knowledge and allegiance. Organizations can leverage the stories to communicate their sustainable brand stories, with a special focus on authenticity in origins and production [24].

Currently, luxury fashion has witnessed a shift toward the production of sustainable products among brands. It means that following the trend of sustainable materials and products in interior design and choosing long-lasting items as exclusive products can meet the growing consumer needs for responsible luxury. Such projects like Gucci's "Vault" store items and pieces that can be considered vintage or at least upcycling, thus responding to the consumers' demand for something extraordinary yet at the same time respecting the circular economy [25].

The resale programs and repair services are slowly becoming key components of sustainability initiatives of luxury brands. Through the resale of products online and repair services, consumers are able to save on costs and at the same time promote sustainability through brands. It also helps to meet the values of luxury consumers and place these brands at the forefront of the circular economy [26].

4.1 Transparency and Authenticity

Transparency plays a significant role in ensuring that consumers are confident in the firms and their sustainability claims. Companies need to disclose information on the origin of their products and services and the methods they use to manufacture their goods, lest they be accused of being mercenary in their ecological appeals. Successful measures that ensure supply chain transparency including collaboration with firms like Oritain enable customers to authenticate statements on raw material sourcing.

5 The Path Forward to Balancing Luxury and Sustainability

The luxury fashion industry has to balance luxury and sustainability because sustainability is becoming increasingly relevant to consumers. Luxury brands must make desirable, sustainable products without losing the high standards and exclusivity that define luxury. Such brands as Gucci and Stella McCartney represent these directions, leveraging eco-friendly materials while preserving beautiful designs appealing to an emerging market of eco-conscious luxury consumers. Best sustainability practices are also developed by saving resources for future generations, encouraging the product's longevity, and improving the environment's health. Circular fashion is one of the most critical directions toward durability, repair, and recyclability, helping minimize waste and extend the life cycle of luxury goods. By focusing on sustainable materials and practices, the brands are ensuring their future and fulfilling the increasing demand for environment-friendly luxury goods [27].

5.1 A Call to Action for

Collaboration across the luxury fashion ecosystem is important to tackle sustainability challenges. Brands, governments, non-governmental organizations, and consumers must come together to create guidelines for sustainable practices. For instance, The Fashion Pact is a good example of a place where 160 brands unite to address issues such as biodiversity, climate change, and ocean conservation. Collective efforts can make a difference, as proven by the success of this pact [26]. These will develop creativity and build a harmonized approach toward tackling environmental issues within the industry. Industry standards need to be established for sustainability. This involves certification, accountability, and international relations with other industries. Certification includes GOTS and Fair Trade, which ensures that the consumer is convinced of how production was conducted [28]. Innovation also includes blockchain, which can inform consumers about origins. Scaling sustainable innovation would be the next step for the luxury industry, wherein large brands will lead the development of sustainable materials and processes. In contrast, smaller brands pick up through partnerships or industry networks [29]. Further development in eco-friendly materials and production processes would then be facilitated through partnerships with technology companies to enable the luxury industry to be environmentally efficient and sustainable.

6 Conclusion

The concept of sustainable development and the flow of the fashion business toward zero waste is becoming increasingly urgent. These changes include sourcing for novel fabrics such as recycled fibers and employing sustainable manufacturing techniques that limit waste. Francesca is a brand that needs to incorporate ways of interacting with consumers through truthful and informative communications with them. Luxury brands, consumers, and policymakers should undertake the responsibility of senior luxury; brands need to make environmental adjustments and consumers need to promote responsible luxury consumption. By doing so, the industry assumes a collective responsibility of creating global sustainability standards. Finally, the transformation of the concept of luxury so that ethical practices are at its core provides work appropriateness and interaction in a world where changes are constant and makes luxury compatible with environmental responsibility.

References

1. Ranfagni, S., & Ozuem, W. (2022). Luxury and sustainability: Technological pathways and potential opportunities. *Sustainability (Switzerland), 14*(9), 1–12. https://doi.org/10.3390/su14095209
2. Kate. (2023, May 11). *How sustainability is reshaping the luxury industry*. Savanta. https://savanta.com/knowledge-centre/view/how-sustainability-is-reshaping-the-luxury-industry/
3. Kaur, H., Choudhary, S., Manoj, A., & Tyagi, M. (2024). Creating a sustainable future: Insights into brand marketing in the luxury fashion industry. *Cogent Business & Management, 11*(1). https://doi.org/10.1080/23311975.2024.2328391
4. Vanita Dhand—Panchal MA Int Mkt and Business. (2024, April 29). *Luxury redefined: The intersection of sustainability and opulence*. https://www.linkedin.com/pulse/luxury-redefined-intersection-sustainability-opulence-vanita-zvffe/
5. Gupta, R., Kushwaha, A., & Dushyant Dave, N. M. (2022). Waste management in fashion and textile industry: Recent advances and trends, life-cycle assessment, and circular economy. In *Emerging trends to approaching zero waste* (pp. 215–242). Elsevier. https://doi.org/10.1016/B978-0-323-85403-0.00004-9
6. Chamorro, A., & Ramírez, R. R. (2023). Green is in fashion: Eco-design and circular economy strategies in the fashion industry. In *Responsible consumption and sustainability* (pp. 49–63). Springer. https://doi.org/10.1007/978-3-031-30742-3
7. Birch, K. (2024, February 19). Top 10: Sustainability strategies of luxury fashion brands. *Sustainability Magazine*. https://sustainabilitymag.com/sustainability/top-10-sustainability-strategies-of-luxury-fashion-brands
8. Browne Jacobson. (2022, October 26). *Luxury brands and sustainability: The challenges and solutions*. https://www.brownejacobson.com/about/news-media/luxury-brands-and-sustainability
9. Krentcil, F. (2024, May 15). *Can luxury fashion ever be fully sustainable?* https://www.bbc.com/culture/article/20231108-can-luxury-fashion-ever-be-fully-sustainable
10. Euromonitor. (2023, November). *Sustainability in luxury and fashion: Time for action*. https://www.euromonitor.com/sustainability-in-luxury-and-fashion-time-for-action/report
11. Made by Field—Digital Design Agency. (2024, September 1). *Sustainability in luxury fashion—How premium brands are embracing sustainable eCommerce practices*. Vervaunt. https://vervaunt.com/sustainability-in-luxury-fashion-how-premium-brands-are-embracing-sustainable-ecommerce-practices
12. Gabardi, C. (2024, June 11). *Fashion and sustainability: How luxury brands are innovating with eco-friendly materials*. Luxonomy. https://luxonomy.net/fashion-and-sustainability-how-luxury-brands-are-innovating-with-eco-friendly-materials/
13. Moore, J. (2022, January 28). *High fashion brands are creating luxury with recycled plastic*. This Is Plastics. https://thisisplastics.com/innovation/high-fashion-brands-are-creating-luxury-with-recycled-plastic/
14. O'Brien, L. (2023, March 12). *10 great fashion designers with the most sustainable clothing materials*. Remeant. https://www.remeant.com/10-great-fashion-designers-with-the-most-sustainable-clothing-materials/
15. Balla, T. (2023, May 2). *How to cut waste with lean manufacturing & on demand print*. Mimeo. https://www.mimeo.com/blog/reduce-waste-lean-manufacturing/
16. Worximity. (n.d.). *Top lean manufacturing principles to reduce waste*. https://www.worximity.com/blog/top-lean-manufacturing-principles-to-reduce-waste
17. Enes, E., & Sayğili, B. (2023). Zero waste and modular fashion design: A suggested collection. *Tekstil ve Mühendis, 30*(131), 180–191. https://doi.org/10.7216/teksmuh.1262844
18. Rago, G. (2023, March 17). *Luxury circularity: How brands are changing the mould of consumption*. Condé Nast College. https://www.condenastcollege.ac.uk/news-events/luxury-circularity-how-brands-are-changing-the-mould-of-consumption/

19. Brabners. (2024, May 21). *Brabners | Luxury brands and the circular economy*. https://www.brabners.com/insights/retail-leisure/luxury-brands-and-the-circular-economy
20. Sarmiento, K. (2023, November 20). *Eco-friendly textile innovations: The future of sustainable fashion*. SIM Social Impact Movement. https://socialimpactmovement.org/eco-friendly-textile-innovations-the-future-of-sustainable-fashion/
21. Cohn, A. S. (2021, July 6). *Luxury brands are embracing circular fashion*. Harper's Bazaar. https://www.harpersbazaar.com/fashion/designers/a36577366/circular-fashion-june-july-2021/
22. TDi Sustainability. (2021, April 9). *The fashion pact*. https://www.tdi-sustainability.com/the-fashion-pact/
23. TDi Sustainability. (n.d.). *The fashion pact*. https://www.tdi-sustainability.com/the-fashion-pact/
24. Fashion Revolution. (n.d.). *The G7 Fashion Pact: What it is and what it's missing*. https://www.fashionrevolution.org/the-g7-fashion-pact-what-it-is-and-what-its-missing/
25. Cole, J. (2024, October 11). *Sustainable luxury fashion: Eco-friendly trends and practices*. BlockApps. https://blockapps.net/blog/sustainable-luxury-fashion-eco-friendly-trends-and-practices/
26. Fibre2Fashion. (2024, May 6). *Speakable content*. https://www.fibre2fashion.com/news/apparel-news/global-fashion-firms-achieve-progress-in-sustainability-goals-report-295053-newsdetails.htm
27. Weinlein, J. (n.d.). *Sustainability in luxury—Balancing opulence with responsibility*. Luxe Beat Magazine. https://luxebeatmag.com/sustainability-in-luxury-balancing-opulence-with-responsibility/
28. OTA. (n.d.). *Global organic textile standard (GOTS)*. https://ota.com/advocacy/organic-standards/fiber-and-textiles/global-organic-textile-standard-gots
29. Label and Narrow Web. (2024, April 8). *Balancing luxury and sustainability in labels and packaging*. https://www.labelandnarrowweb.com/contents/view_experts-opinion/2024-04-08/balancing-luxury-and-sustainability-in-labels-and-packaging/

Index

Sadhna et al. (eds.), *Crafting Sustainability in Luxury Textiles for a Zero-Waste Future*, Sustainable Textiles: Production, Processing, Manufacturing & Chemistry, https://doi.org/10.1007/978-3-031-86773-6

The manufacturer's authorised representative in the EU is Springer Nature Customer Service Centre GmbH, Europaplatz 3, 69115 Heidelberg, Germany. If you have any concerns regarding our products, please contact ProductSafety@springernature.com

Printed and bound by CPI Group (UK) Ltd, Croydon, CR0 4YY
30/09/2026
02225563-0002